Facilities and Workplace Design

An Illustrated Guide

by

Quarterman Lee

with
Arild Eng Amundsen
William Nelson
Herbert Tuttle

Engineering & Management Press
Institute of Industrial Engineers
Norcross, Georgia, USA
http://www.iienet.org

Published by the Institute of Industrial Engineers.
Printed in the United States of America.

02 01 00 99 98 97 6 5 4 3 2 1

CATALOGING-IN-PUBLICATION DATA

Lee, Quarterman.

Facilities and workplace design: an illustrated guide/by Quarterman Lee; with Arild Eng Amundsen, William Nelson, Herbert Tuttle.

p. cm. -- (Engineers in business series; 3)

Includes bibliographical references and index.

ISBN 0-89806-166-0

1. Plant layout. 2. Plant engineering. 3. Office layout. 4. Work environment. I. Amundsen, Arild Eng. II. Nelson, William. III. Tuttle, Herbert. IV. Title. V. Series.

TS178.L44 1996

658.2'3--dc20 96-22972

CIP

Director/Engineering & Management Press: Eric E. Torrey
Editor: Forsyth Alexander
Cover by Marty Benoit
Printed by Edwards Brothers Inc.

ISBN 0-89806-166-0

Engineering & Management Press
25 Technology Park
Norcross, GA 30092

Table of Contents

Preface

Modern factory layout began most notably at Highland Park, Michigan. There, Henry Ford and Charles Sorensen first put large-scale flow line production into place. Layout, the arrangement of equipment within the building space, was the most striking feature of Ford production. The layout of Ford factories was so noticeable, it disguised the more intangible elements of Ford and Sorensen's highly integrated and well-thought-out manufacturing system. As a result, many factory managers, business executives, and engineers still try to replicate early Ford layouts. These imitations often are ill-suited to their business purposes. Even when managers do not copy Ford layouts, they continue to believe that layout is the heart of efficient production.

And so it is, but only as the culmination of a rational design process: a process that moves from global location to workstations, from policy to operations. It is a process that includes the entire range of tangible and intangible elements. The result is a design that integrates products, service, people, information, and technology.

Our facilities planning approach builds on the pioneering work of Richard Muther and Knut Haganas, both of whom developed systematic layout planning (SLP) and systematic handling analysis (SHA). These concepts remain valid after almost thirty-five years. Our approach uses contemporary language and examples. Data acquisition is stressed. Perhaps most importantly, we integrate manufacturing strategy and manufacturing focus, concepts first put forth by Wickham Skinner and first recognized as connected by William Wrennall.

This is more than a manufacturing book, however. The concepts originally developed for industry now apply to services. Products and services meld together in almost every viable enterprise in today's world. The concept of "business process" covers all productive human activity. Government, education, the professions, and industry are simply variations on a few central ideas. Therefore, the planning of a government facility or an idea factory follows the same principles and approach as the planning of an automotive factory. The planning of a hospital follows the same course as the planning of a textile mill.

A facility project often is the catalyst for new strategies. Under its guise, management may open a strategic debate that brings together the many disparate functions and elements of the enterprise. It can provide the structure for analysis and discussion. The project then translates the results into steel and concrete. In this way, facility planning becomes a large-scale reengineering project rather than an exercise in template shuffling.

Our purpose in writing this book is to help practicing industrial engineers new to facility planning. We intend it to be practical with many examples, forms, diagrams, and visual aids. Visual approaches are especially important for facility planning. The human mind is usually more creative in a visual mode. Illustrations, graphs, and charts effectively represent complex spatial arrangements that have many levels of detail. Also, data is more significant for most of us when we can see it.

A plant layout is the product of thousands of decisions, both past and present. It is the physical manifestation of the firm's manufacturing strategy, whether de facto or otherwise. These decisions cover the entire range of manufacturing—finance, personnel, process, product design, and many, many other topics. We cannot possibly do justice to all of these issues. We hope, however, to alert industrial engineers to their existence and importance. We hope to point the way to more rational and strategically oriented facility design.

Acknowledgments

We take this opportunity to thank others who helped with this book. Margie and Curt Jennings produced much of the artwork. Sandra Lee helped with research, data entry and proofing. Maura Reeves, Ellen Snodgrass, Eric Torrey, Forsyth Alexander, and the staff at the Institute of Industrial Engineers have supported us and helped make this final product readable.

Chapter 1

Facilities in a Changing Environment

"The dogmas of our quiet past are inadequate to the stormy present. As our situation is new, we must think anew."

— Abraham Lincoln

Working facilities are the land, buildings, and equipment that provide the physical capability to add value. This book is about operational facilities used for a wide range of business, government, institutional, and charitable activities. It applies to offices, factories, and fast-food restaurants. It applies to any facility that houses value-adding operations. For convenience, terms such as "business facility" or "factory" are used, although the changing nature of work has blurred many of these distinctions. The principles herein apply to a wide range of situations the industrial engineer commonly encounters.

Facilities are both durable and expensive, lasting for decades and sometimes even spanning centuries. A firm's facilities are among the most expensive of its possessions. They represent the largest asset item on most balance sheets.

The durability of facilities, their cost, and their primary role in adding value make them an important strategic element. Just as gunpowder made the fortresses of medieval Europe indefensible, changes in technology, culture, and politics can quickly render today's industrial facilities obsolete. Conversely, facilities that adapt to the nature of their competitive environment can be a continuing source of advantage for their owners.

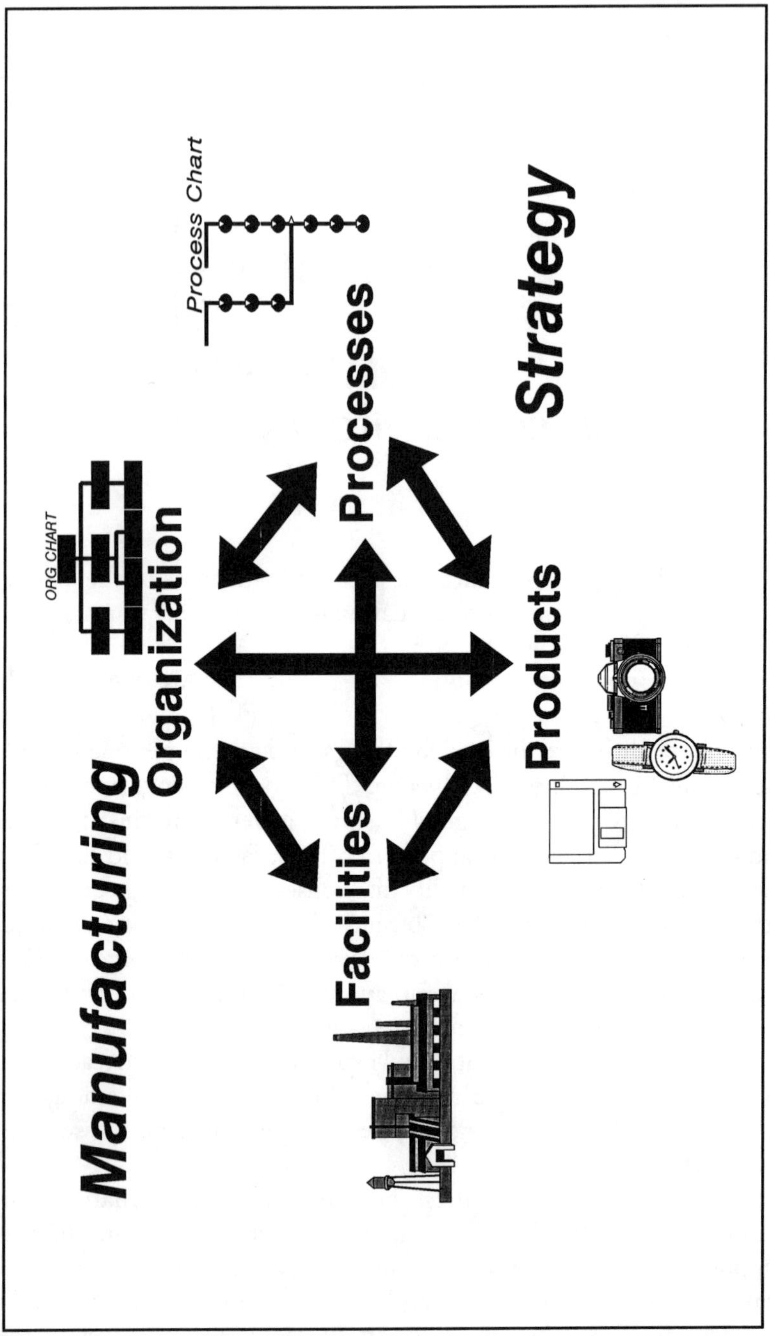

Figure 1.1 - Strategic Relationships

Figure 1.1 depicts the interaction of facilities, organization, products, and processes. The understanding, design, and development of these varied elements into a functioning business system are referred to in various terms. Among these are: manufacturing strategy, corporate reengineering, and business architecture.

The importance of facilities does not lie solely in their cost and durability. They are also the most tangible element of the business system, the element to which everyone in every area of the business can relate. They can be a central, common reference for the restructuring/reengineering/strategic debate.

Working facilities in modern history

Industrial facilities

Shops that served the needs of individual artisans were the industrial facilities of the Middle Ages. These were small and centered around a single skill such as armory or saddlemaking. They had simple and clear arrangements.

During the Industrial Revolution, power sources and the movement of raw materials determined facility design. Textile mills required streams for water power, and cumbersome shafts and belts dominated their arrangement. Early iron and steel mills were located on waterways, railroads, or mining sites; coal, iron ore, and limestone transportation dominated their design.

Early large-scale production shops such as the Pickering Piano Factory (fig. 1.2) developed in the nineteenth century. These large

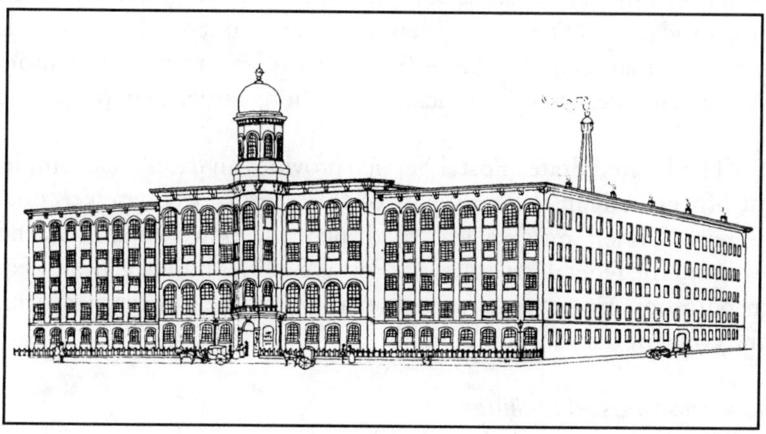

Figure 1.2 - The Pickering Piano Factory, Boston, Mass, Circa 1870

buildings turned out high numbers of manufactured products. At one time, the Pickering factory turned out 400 pianos each day.

In the early twentieth century, the progression of mass-production technology required facilities that optimized material flow. The micro-division of labor made skill less important than efficient movement of product.

In the second half of this century, information and knowledge began to dominate industrial production. The education and skills of the workforce in industrialized parts of the world increased. As a result, industrial facilities must now optimize the coordination of people, processes, and products.

Government facilities

In the Middle Ages, the most important government facilities were town fortresses. Their primary mission was defense against roving bands and neighboring city-states. The fortified town of Rocroi, on the northern plain of France, is an example. Still largely intact, it is a lasting testimony to the durability, cost, and obsolescence of these fortresses.

With the advent of gunpowder, battle technology advanced. New tactics evolved and armies became more disciplined. These massive works drained the treasuries of many dukes and kings and became indefensible and obsolete. By the time of the Renaissance, fortresses had evolved into palaces. Their primary mission was comfort for the inhabitants, as well as the projection of power and prestige. The builders of many governmental buildings wanted to intimidate potential enemies, both foreign and domestic.

Governments no longer can survive only through warfare or the threat of warfare. Their constituents demand added value in a wide range of human activity. Accordingly, many governmental facilities now are being designed for efficient operations rather than projection of power.

The United States Postal Service provides an excellent example. Post offices built in the early part of this century were architectural landmarks. Their mission was to display the power, stability, and prestige of the federal government. Postal facilities built today are near transportation centers and optimize mail flow. Their primary mission is the efficient distribution of mail.

Knowledge-based facilities

Facilities in which knowledge is the primary means of work have always been more varied than other types. The medieval monastery, for example, was a primary depository of knowledge in its time.

The church used this knowledge to vie with governments for power and influence.

During the Renaissance and Industrial Revolution, knowledge became an important source for commercial competitive advantage. Individual professionals such as doctors, lawyers, and financiers were primary keepers of knowledge. Other knowledge resided in libraries. Factories imbedded it in their facilities and processes. Peter F. Drucker was among the first to recognize the increasing value of what he termed "knowledge work." He put forth these ideas in his landmark work, *The Practice of Management,* in 1955. Knowledge work depends primarily on brainpower rather than manual skills or strength. In today's manufacturing environment, most work requiring pure strength of muscle has long been automated away. Much of the work that once required manual dexterity has been taken over by computerized equipment such as numerically controlled machine tools or coordinate measuring machines. Therefore, knowledge and the information behind it now have become primary sources of value in their own right. Many organizations exist for the sole purpose of processing information and distributing it. Their facilities should reflect and enhance this role.

Facilities in a changing environment

Facility designers have always worked with materials, products, processes, information, and people. Their task is to arrange work processes on land and in buildings for optimum performance. This has not changed and will not change, but rapid shifts in technology, politics, and culture require a more fundamental understanding and analysis from the facility designer. It no longer is sufficient (if it ever was) to copy an assembly line just because it was successful somewhere else.

In addition to the long-term trend toward increased knowledge-based work, other trends of a stretigic nature are affecting business. The facility planner should catalyze or lead an organization's adaptation to ever-changing surroundings.

The environmental imperative

Harmony with the environment is an increasingly important business concern that will not go away. Population growth is a principal factor dictating this concern; the spread of the suburbs through increased mobility is another. Organizations that survive and prosper in coming years will anticipate and lead with their environmental policies.

Location requirement changes

Information is the raw material of the knowledge worker. With the confluence of information processing and communication, the information superhighway has opened. The ability to distribute vast amounts of information makes it less important for facilities to locate near the source of information. This is similar to the distribution of materials in an earlier day. As material transportation became more efficient, manufacturers could locate farther from their sources.

Knowledge-based facilities now locate where their workers wish to live—often far from traditional industry. Industries that require specialized knowledge often converge in small areas: manufacturers of overhead cranes congregate in Milwaukee, Wisconsin; Wichita, Kansas, has a high concentration of vinyl printers for decorative decals; and "Silicon Valley" in California is the home for many electronics plants. These changes affect global facilities planning decisions such as site selection and planning.

The changing nature of work

As products become more sophisticated, their knowledge component becomes more important. It is no longer enough to manufacture a commodity product. Competition demands variety, frequent change, and distribution systems that deliver physical product, service, and knowledge.

The nature of work has changed. Today, individuals seldom work alone. Knowledge teams are necessary in product design, process design, finance, and even law. Teams, by their nature, require proximity. Facilities can inhibit or promote teamwork. They can smooth the operation of complex and ethereal knowledge processes or they can isolate people and prevent communication.

The socio-technical system

Socio-technical systems have always existed, although few managers recognized the phenomenon until recently. Management thought was caught in the Newtonian concept that organizations were like machines, giant clockwork mechanisms that ticked away in a predictable, mechanical manner. Eric Trist of the Tavistock Institute developed the socio-technical idea in the early 1950s. Teamwork, total quality management (TQM), and other techniques for employee involvement have their roots in the concept of the socio-technical system (fig. 1.3).

The social system includes people and their habitual attitudes, values, behavioral styles, and relationships. It is the formal power structure depicted on organization charts and the informal structure derived from knowledge and personal influence. The technical system includes machinery, processes, procedures, and their physical arrangement (layout).

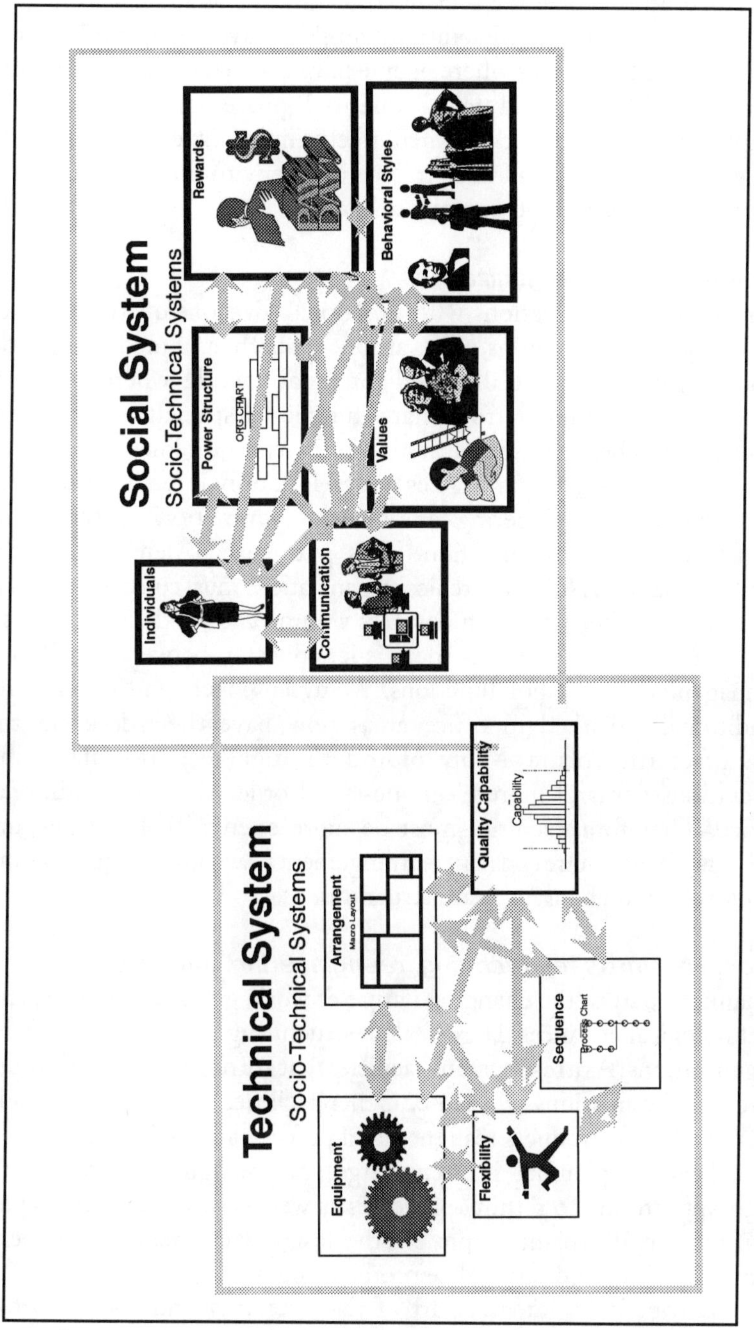

Figure 1.3 - Socio-Technical Systems (Eric Trist, 1952)

To be effective, the social and technical systems must integrate and assist one another. Facilities planning plays a major role in this integration. Businesses where people have isolated workstations, large inventory buffers, and few sequential processes have difficulty implementing teamwork. A manufacturing work cell that requires extensive teamwork will not produce in an environment of suspicion, individual rewards, and command-control.

Non-hierarchical organizations

Hierarchical organizations with functional divisions of work evolved from the Roman Legions, the Catholic Church, and medieval guilds. Such organizations are ill-suited for today's work, where the work product requires input from many functional specialties and where coordination between specialties is a primary requirement.

While TQM emphasizes cross-functional teamwork, more fundamental reengineering emphasizes elimination of functional structures in the organization. This puts special demands on the facilities planner. Non-hierarchical organizations must constantly change to accommodate changes in business volume and product life cycles.

In these organizations there is less division between traditional management and labor functions. Many engineers and others who traditionally worked in office areas now have their desks in the manufacturing plant. Many of today's high-tech manufacturing operations demand more cleanliness and order than the traditional office. Therefore, facilities must be more open with few walls and barriers. They require constant rearrangement to accommodate changing work cells and changing team structures.

Global business restructuring, reengineering, and facilities

Thanks in part to the changing nature of work, global economics, and technological advances, large-scale restructuring is occurring in many organizations. As a result, many facilities that are no longer contributing to company missions will close. Other facilities will be built. Many more will have products realigned and processes reengineered.

Facilities planning is often a large-scale reengineering project. It is an opportunity to rethink processes as well as supporting elements. During a facilities planning project, the designers can help management clarify missions and rationalize product lines.

Layout is an integral part of reengineering and restructuring. Meaningful restructuring requires corresponding changes in the layout.

Conversely, a layout redesign can be the catalyst for restructuring. Many symptoms of inappropriate business architecture appear as layout or material handling issues. Factory layout can demonstrate the need for reengineering to an organization reluctant to tear itself apart and rebuild.

Approaches to facility planning

Those who plan and build facilities take many approaches. Some are highly organized; others are ad hoc. Examples of approaches (fig. 1.4) are *experiential*, *master building*, *cloning*, *bottom-up*, *systematic*, and *strategic*.

Experiential

In this approach, people plan their facilities based on past experience, common sense, and instinct. In any organization, the experience of senior members is valuable for information on what has worked and what has not worked in the past. Organizations, as well as individuals, need this experience to function.

A facility designed from experience taps into the rich knowledge of those who have gone before; however, experience-based facilities planning has limitations. Experience, by definition, is based on the past, and new technology and organization structures can make it obsolete. In addition, planning by experience is usually unorganized. It frequently is the result of the memories of only one or a few individuals, and others may have had additional or contradictory experiences. Such hindrances, as well as forgotten details, haunt these efforts.

In planning a major facility, experience cannot be ignored but must be gathered from the widest field of experience possible and applied with judgment and discretion.

Master building

Master building focuses on construction and buildings. The final product is often impressive and sometimes a work of art, but it may not fit the operational needs of the enterprise. Master builders can be found at many levels in both large and small organizations: a company president building a new headquarters or a department manager focusing on technological impressiveness rather than actual needs. Using a building to display financial strength, technological prowess, or artistic accomplishment is a legitimate form of advertising. However, this purpose should be balanced with other business needs.

Cloning

Cloning simply duplicates an existing facility or portion of it. This approach is fast. If the existing facility is proven and if conditions are the same, this type works well. McDonald's uses cloning to build its hamburger "factories" throughout the world. For most facilities, however,

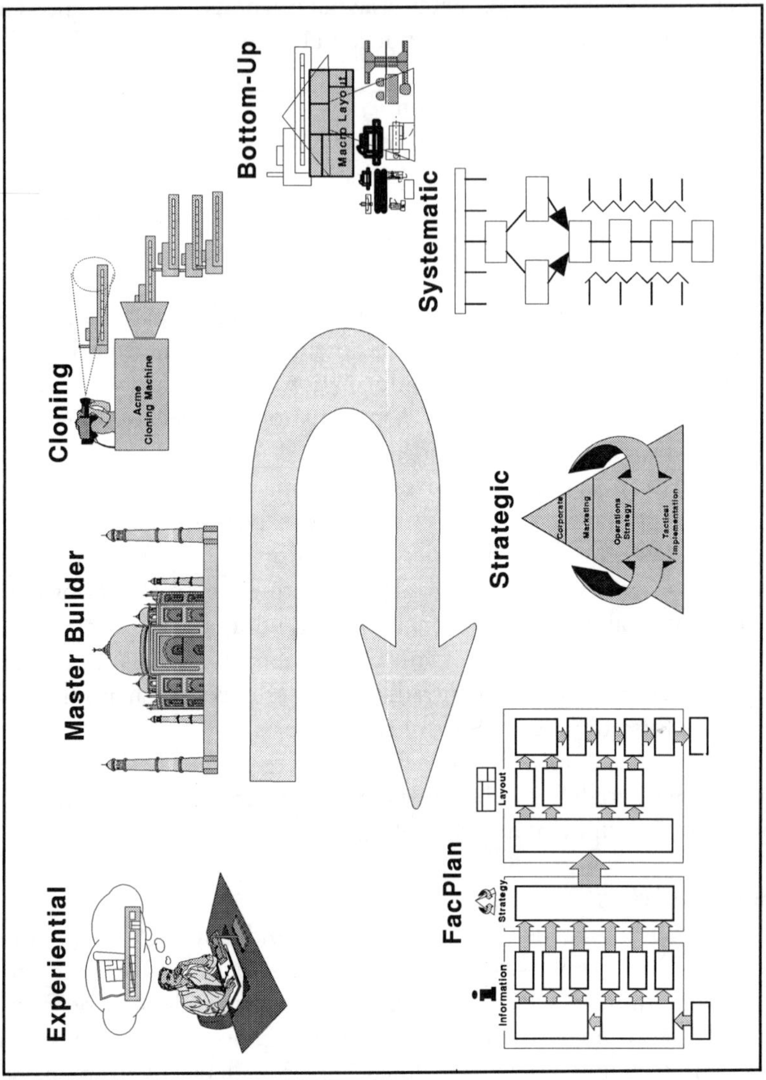

Figure 1.4 - Spaceplanning Approaches

cloning has limited use because sites, processes, and people are different. Cloning should be applied only when appropriate.

Bottom up

The bottom-up approach starts with the details. How many desks? How many and which machines? How many people? From them, departmental units and, eventually, the overall facility plan are built. It is a satisfactory approach if the details and how they will be assembled into a larger system are known, if there is time, and if the details will not change. Such conditions are often met for smaller facilities in stable environments.

Bottom-up planning does not lend itself to new operations strategies. Because all details have to be worked out before final design and construction, construction lead times are often too long. On large projects, the details become so overwhelming it is often difficult to maintain schedules.

Systematic

Systematic layout planning (SLP) uses procedures, conventions, and phases. It helps layout planners know what to do at each step of a project. This provides layout planning with system and structure, saving time and effort. However, many layouts created with systematic methodology are simply better versions of what went before. The primary concern is how to arrange blocks of space. A more fundamental issue is what blocks of space should be arranged.

Strategic

The strategic approach is top-down. It sets policy first and arranges the technology, organization, and facilities to support it. Starting with business and corporate strategy such as global site location, it moves to operations strategy and finishes with details like locations of equipment and furniture.

A strategic approach is direct and has purpose. It allows everyone involved in the project to follow a common direction. Used alone, however, strategic direction is insufficient. It does not tell facility designers and those who use the facilities what to do.

FacPlan

The FacPlan method combines the best of various approaches. It has system and structure and adds strategic dimension. It taps into the experience and knowledge of those who use the facilities. It can work from detail to general and vice versa when appropriate.

FacPlan uses a hierarchy of detail levels. It focuses on strategic issues at the appropriate time and minutiae at the appropriate time, using a model project plan to guide and structure each project. Procedural flow charts guide the planner through each task and assist with decision making. Charts, forms, and design aids contribute to the organization of information.

The industrial engineer's role in facility planning

The central, strategic role of facilities places their designers in a unique position. Industrial engineers can assume narrow roles as technical equipment arrangers or they can take broader roles as educators and catalysts for organizational strategic debate.

The latter requires more than skills in layout design and technical procedure. Strategic perspective, well-developed interpersonal skills, patience, and understanding are also necessary.

This work provides insight into the basic technical tools industrial engineers need for facilities planning. The broader skills will require experience, insight, maturity, and education.

Chapter 2

The Framework for Facilities Design

The complete design of a facility requires work from many disciplines within an organization: sales and marketing, purchasing, human resources, accounting, and more. More visible is the work of architects, structural engineers, process engineers, and management. Architects and structural engineers check soil conditions, building codes, and infrastructure, detailing the structure, appearance, and internals of the building and site. Process engineers may plan the production procedures. To guide and coordinate all these efforts, management sets strategic policies.

Industrial engineers also play key roles. They often manage the overall project and report to top management, and they may perform some or all of the above tasks. Most importantly, they plan the use of space. These space plans, at various detail levels, become the centerpiece for coordinating the entire project.

The levels of spatial design

Layout, or space planning, is the central focus of facilities design and dominates the thoughts of most managers. But factory or office layout is only one detail level. Ideally, a facility design proceeds from the general to the particular—from global site location to workstation. Larger strategic issues are decided first.

It is useful to think of space planning in five levels as shown in figure 2.1. Figures 2.2 through 2.6 show typical outputs at each level. These range from the global maps of site location to engineering drawings of tools and workstations.

Level	Activity	Typical SPU	Environment	Output
I Global	Site Location & Selection	Sites	World Or Country	
II Supra	Site Planning	Buildings Or Site Features	Site	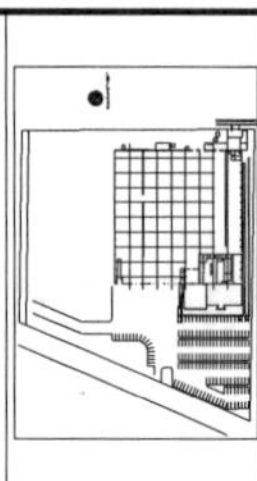
III Macro	Building Layout	Cells Or Departments	Building	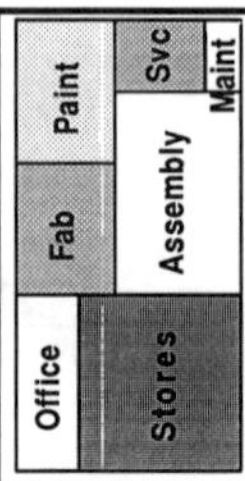
IV Micro	Department Or Cell Layout	Workstations Or Cell Features	Cells Or Departments	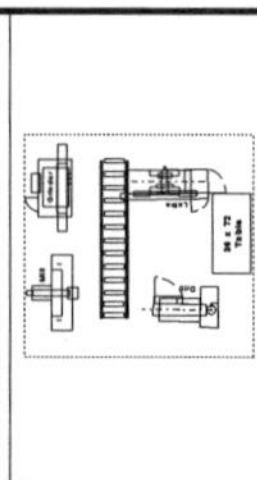
V Sub-Micro	Workstation Design	Tool Locations	Workstation	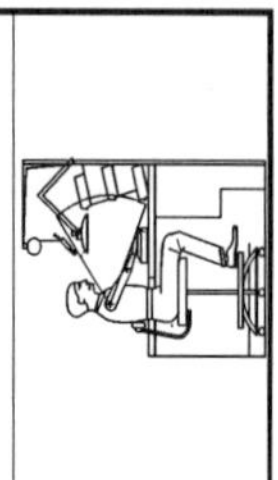

Figure 2.1 - The Levels of Space Planning

Level 1—Global site location

During global location, the site location level, the firm decides where to locate facilities and determines their missions. A facility mission statement is a concise summary of products, processes, and key manufacturing tasks. A facility rarely can perform more than two or three key manufacturing tasks well. The mission statement is therefore an important guide for facilities planners and others as they consider various design trade-offs.

Other outputs at this level usually include a report to management. For multiple sites, maps showing site locations and customer activity are common. Figure 2.2 illustrates.

The cost of space planning at Level 1 is small. Global location usually involves a few top executives and one or two industrial engineers or consultants. Each level below requires more and more people, analysis, and detailed engineering. Yet, the corporate budget process frequently demands that all significant planning be delayed until after a decision is made to proceed with site acquisition. Those levels with the

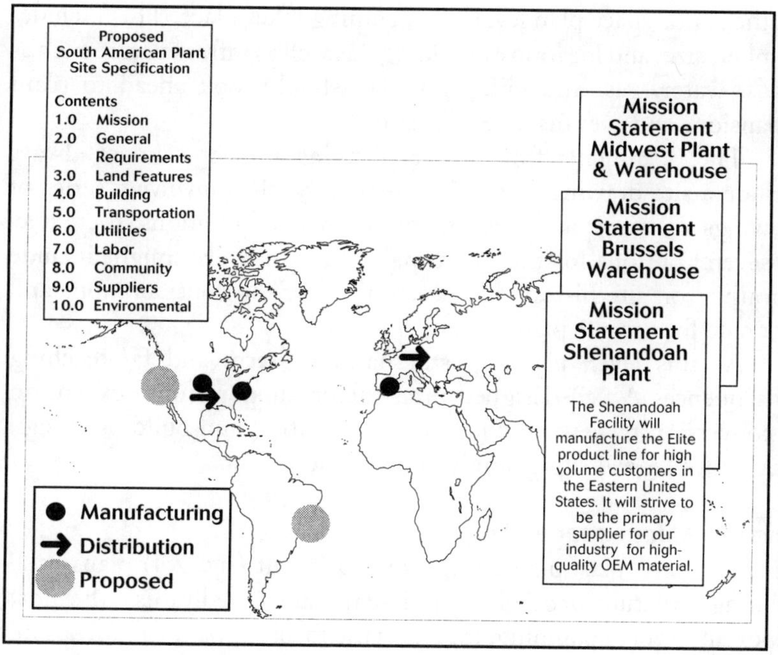

Figure 2.2 - Level I - Site Location

most strategic impact and the lowest planning cost receive the least attention. Consequently, the decisions with the most strategic impact are sometimes made with the least reliable knowledge.

Overall business strategy is most important at the global level. Determining the number and location of sites requires far more than simply searching for the lowest labor rates and largest tax breaks. Available labor skills and attitudes toward work, supporting services such as tool production and material supply, and politics, and sometimes geopolitics, must also be major considerations. For example, if a plant is located in the wrong country, it may become a geopolitical pawn. Technological prowess could then shift to other regions. If there is political instability locally, it can destroy a firm's ability to produce. Important raw materials might be depleted or replaced. Such problems are not easy to correct.

Appropriate planning results in facilities optimized for the markets and located near the most important resources—resources that, increasingly, involve knowledge, skills, and infrastructure rather than raw materials.

Level 2—Supra-space plan

At the supra-space plan level, site planning takes place. This includes number, size, and location of buildings, as well as infrastructure such as roads, water, gas, and rail. This plan should look ahead to plant expansions and eventual site saturation.

The documents from a site planning project almost always include a site drawing (fig. 2.3). Frequently, they involve a series of drawings showing past, present, and future configurations (there may be several options for these). A major site study also might include narratives on site history and descriptions of the considerations and rationale for the site plans.

At this level, planning still has long-term and far-reaching consequences. A well-designed infrastructure supports future expansion or conversion to new products. Proper location and building design provide for logical expansion in suitable increments.

Level 3—Macro-space plan

At the macro-space plan level, a macro-layout (fig. 2.4) plans each building, structure, or other sub-unit of the site. Usually this is the most important level of planning, for it sets the focus, or basic organization, of the factory. The designers define and locate operating departments and determine overall material flow.

Macro-space plan decisions may result in new-product flexibility,

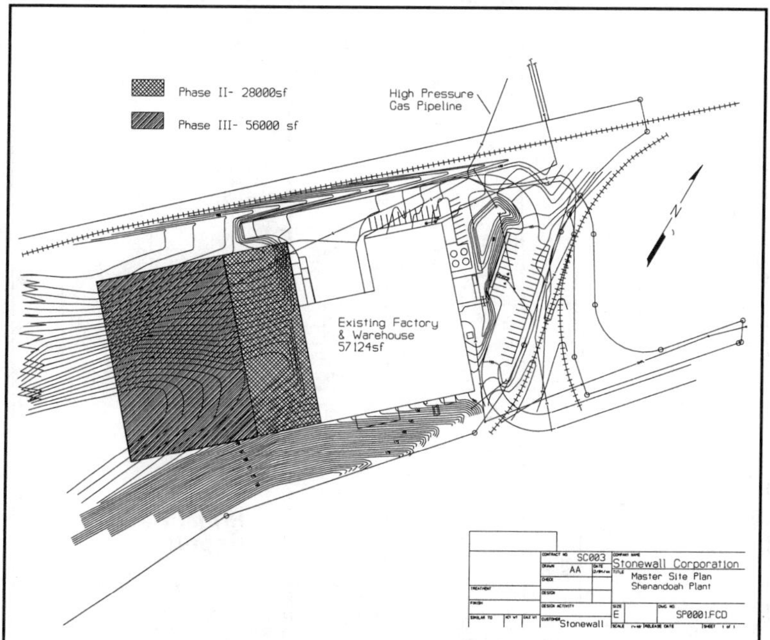

Figure 2.3 - Level 2, Site Planning
Master Site Plan, Shenandoah Plant, Stonewall Corp.

lower costs, high quality, or a flexible labor. Fundamental macro-space plan decisions usually are easier to correct than site-level decisions. Still, a poorly planned facility can bring high handling costs, confusion, and inflexibility. These problems, in turn, can cause difficulty in launching new products, erratic deliveries, and too much inventory. Correcting such problems may require a complete rearrangement with major investments in process equipment and infrastructure.

Level 4—Micro-space plan

The location of specific equipment and furniture is determined in the micro-space plan. The emphasis shifts from gross material flow to personal space and communication. Socio-technical considerations dominate. If production teams are an important element of the operations strategy, the work at this level may inhibit or discourage teamwork. Figure 2.5 shows a space plan for an operating department.

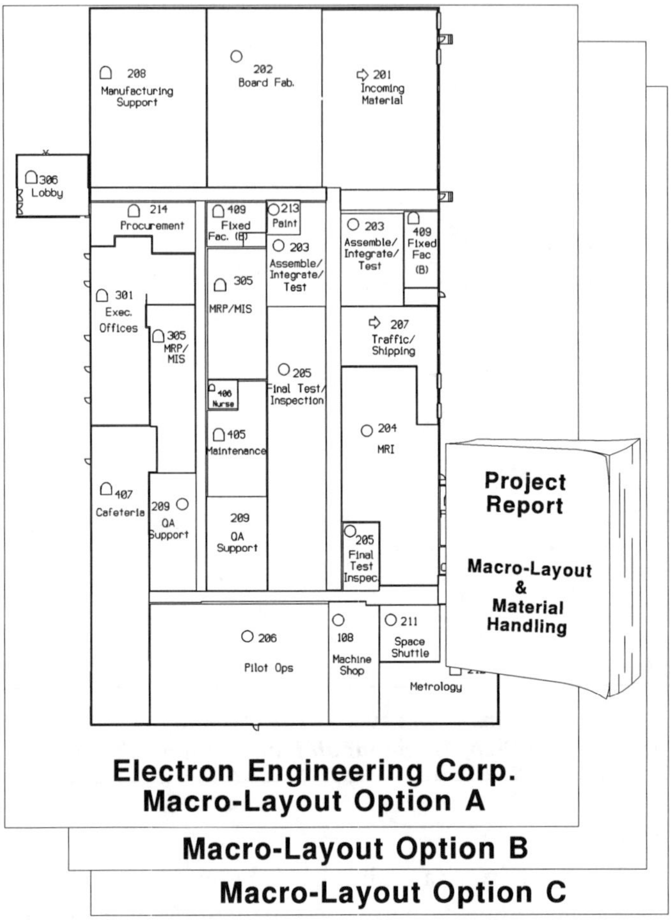

Figure 2.4 - Level 3, Macro Layout
Project Report: Macro-Layout and Material Handling

Level 5—Sub-micro-space plan

Individual workstations and workers are the concern of the fifth level. Here, workstations are designed for efficiency, effectiveness, and safety. Ideally, the industrial engineer plans for the correct tools in the most appropriate places, using fixtures that properly hold the work piece. Materials are introduced at optimal locations and large items are provided with appropriate material handling aids. Some typical outputs are shown in figure 2.6.

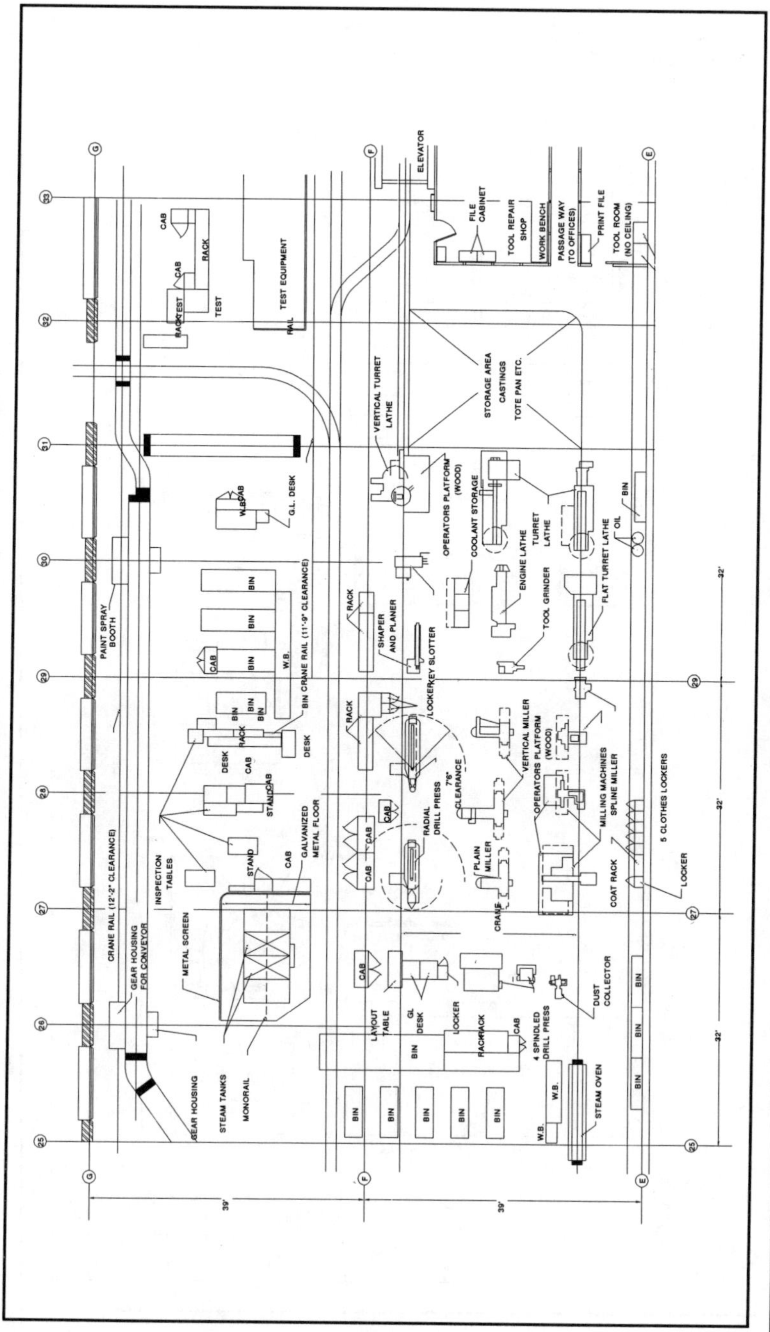

Figure 2.5 - Level 4, Micro-Layout
Maypro Manufacturing Pump Machining Dept. Detail Layout

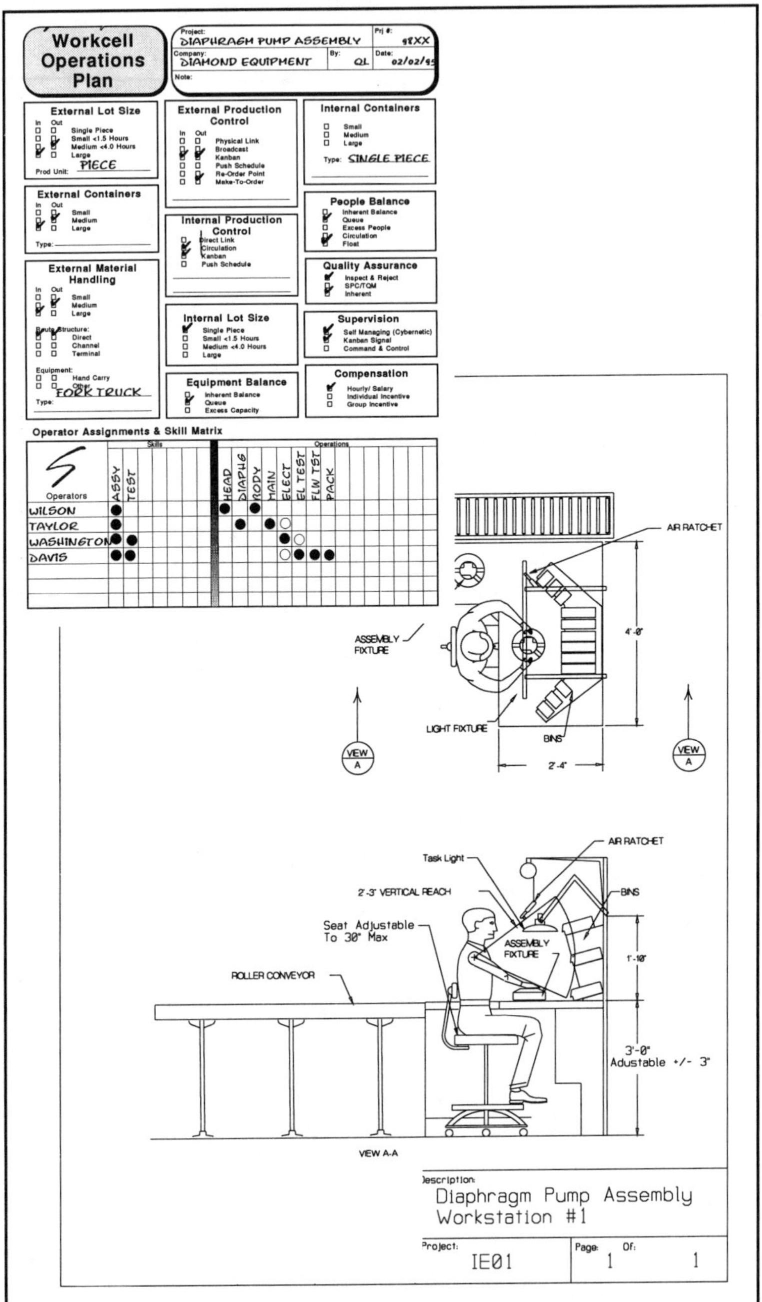

Figure 2.6 - Level 5, Workstation Layout

Levels 4 and 5 are the more detailed levels of space planning; therefore, equipment and issues are more localized. When changes are necessary, there is usually less danger of major production interruptions.

The phasing of space design

Ideally, design progresses from the global level to the sub-micro level in distinct, sequential phases. At the end of each phase, the design is "frozen" by consensus. This settles the more global issues first and allows smooth progress without continually revisiting unresolved issues. It also prevents details from overwhelming the project. Figure 2.7(A) illustrates this logical progression and shows the strategic impact of the work in each phase. Strategic impact affects the long-term ability of the firm to compete and profit.

Industrial engineers rarely have the opportunity to design a facility in accordance with the normal phasing shown in figure 2.7(A). There are several reasons for this. Sites and buildings that have evolved over many years outlive technologies and their original purpose, and therefore must be rearranged. Another reason may be management's belief that the existing space plan is simply not optimal. In both cases, planning

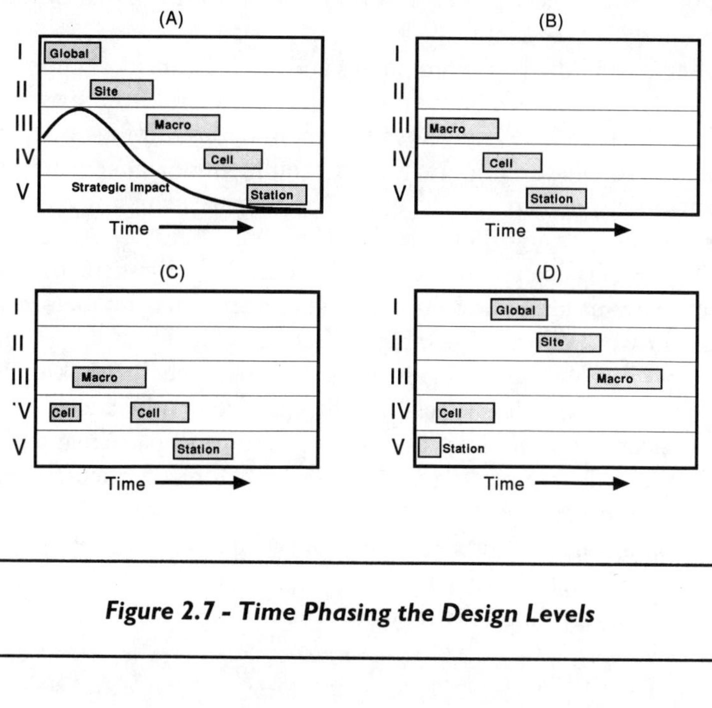

Figure 2.7 - Time Phasing the Design Levels

begins at the macro-space plan level. Figure 2.7(B) illustrates this. The phasing demonstrated in figure 2.7(B) also occurs when management makes global and site-level decisions without the benefit of advice and counsel from their facilities planner(s).

The size and organization structure of cells in a macro-space plan may be indeterminable when processes and strategies are untried. This often happens when firms make a transition from functional to cellular manufacturing. Pilot cells must then be developed to prove the concept or technology. Figure 2.7(C) reflects this. A cell or micro-space plan (Level 4) then becomes the first phase. Upon completion of this pilot, people can agree on the general approach. Then the designer can shift back to Level 3 and prepare a macro-space plan. The details of remaining cells are defined in their optimal sequence.

The phasing demonstrated in figure 2.7(D) is common for large office layout projects. First, the details of workstation layout are established. This may come from standardizing space and equipment based on each person's position in a hierarchy. Secretaries, for example, may get a 175-square-foot workstation with filing space and word processing equipment, while a Grade I engineer gets a 110-square-foot cubicle and a supervisor, a 150-square-foot cubicle. From the organization charts and staffing forecasts, the space for each department and the arrangement between departments can then be developed. At this point, the project moves upward in detail to the global or, more commonly, macro- level.

Separating the work into phases and levels is the ideal approach. Nevertheless, there may be some overlap. For example, the space plan of a particular work cell may not fit the boundaries previously decided in the macro phase. This may then require minor changes to the previously designed and agreed upon macro-space plan. For these and other reasons, phasing should be flexible.

Proper phasing should be considered in the earliest stages of the project, perhaps after the initial discussions and certainly before any significant work effort begins. Here are some guidelines:

- work from the most general to the most specific level (highest to lowest) unless special conditions dictate otherwise;
- clearly communicate the phasing plan to all participants;
- resist the temptation to jump ahead before a particular phase is complete;
- obtain agreement on the plan for each phase before moving on to the next phase; and
- recognize that there may be some overlap between phases.

The space plan elements

Every space plan at each level has four fundamental elements and two derived elements. The fundamental elements are: *space planning units (SPUs)*, *affinities*, *space*, and *constraints*. When developing a space plan, the designers first define and identify SPUs. They then evaluate affinities. Using the affinities, they join SPUs to form one or more *affinity diagrams*. The affinity, or configuration, diagram is the first of the derived elements. Space added to the configuration diagram produces a *space plan primitive*, the second derived element. Constraints applied to the space plan primitive produce the space plan. Figure 2.8 shows this progression.

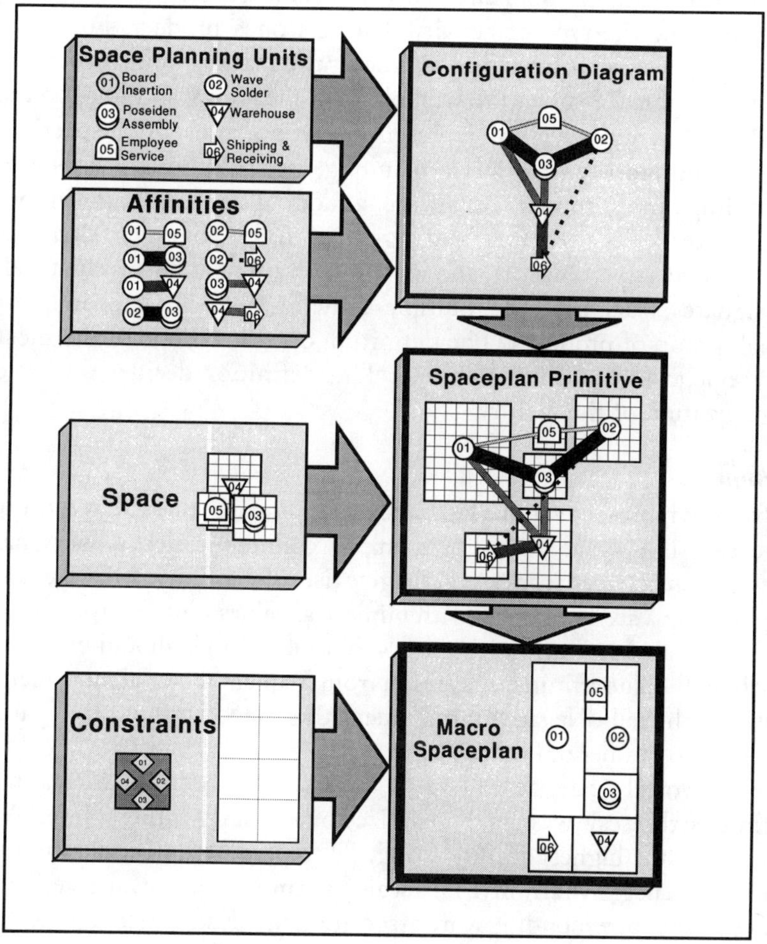

Figure 2.8 - Elements of a Spaceplan

The concept of fundamental and derived elements is valid at all levels. However, it is most useful and direct at the macro- and site levels. The chapters that follow explore its application.

Space planning units

SPUs are the entities arranged by space plan designers. At the macro-level, they are referred to as *cells.* (The systematic layout planning [SLP] system used the term *activity area.*) A cell might be a work department, a storage space, a building feature, or a fixed item. Each cell initially is represented by a symbol and identifier.

Most of these symbols are taken from ANSI Y15.3M-1979, the American National Standards Institute standard for process charts, which show the type of activity that acts on a product. For space planning, the symbol that best represents the space's dominant activity is used. Figure 2.8 shows the symbols, their meanings, and color codes.

The standard symbols represent operation, transport, inspection, delay, and storage. For space planning, two additional symbols—handling and product cells—are added. The handling symbol designates areas used for repackaging, transfers, or other elements that are partly transport and partly operation. The product cell designates space used for multiple activities on a single product or small group of products. The definition of SPUs is one of the most strategic tasks in facility planning. This definition decides the basic organization of the factory.

Affinities

Affinities represent various factors that demand closeness between any two cells in a space plan. For example, communication or personal interaction between workers might give rise to an affinity. Affinities are rated using a six-level scale, with numerical values ranging from +4 to -1. The scale has four positive levels that mean SPUs should be close. Such high-value affinities may result from frequent material movement between the cells. Negative ratings mean that the SPUs should be apart. There also is a neutral rating, 0.

A vowel scale, A-E-I-O-U-X, may also be used for rating affinities; this scale was first popularized by Richard Muther. Here, "A" represents the highest affinity rating, "U" represents a neutral affinity, and "X" is a negative affinity. This scale has a mnemonic advantage. The vowels have corresponding word associations as illustrated in figure 2.9. Chapter 3 discusses the methods for evaluating affinities.

Figure 2.9 shows the affinity conventions developed by building on the original SLP system. The multi-line representation works well

Description	Vowel Rating	Scalar Rating	Manual Graphic	CAD Graphic	Color
Absolute	A	4			Red
Exceptional	E	3			Yellow
Important	I	2			Green
Ordinary	O	1			Blue
Unimportant	U	0	(None)	(None)	(None)
Apart	X	N/A			Black

Figure 2.9 - Affinity Conventions

for manual graphics. On many CAD systems and other computer graphics software, it is easier to use varying line widths, gray scales, and color. When color is available, it dramatically illustrates the nature of the affinity network.

Figure 2.10 shows the typical range of affinity distributions for macro- and micro-layouts.

Affinity diagram

SPUs combine with affinities to form an affinity diagram—the first of the derived elements. This diagram is an idealized spatial arrangement that eventually becomes a space plan. In the diagram, symbols represent SPUs and lines represent affinities between them. A single line is the lowest value affinity and a four-part line is the highest. Squiggly lines represent negative affinities. These conventions are illustrated in figure 2.9.

Using an iterative process, the designer manipulates the diagram to create an optimal or near-optimal arrangement. A near-optimal arrangement has very short high value affinities at the expense of lower value affinities. It minimizes the crossing of affinity lines

Figure 2.11 illustrates the iterative improvement of an affinity diagram. It is interesting that many computerized planning systems emphasize this specific process when, in fact, it is the part of the layout process to which computers are least suited.

Space

Each SPU has a unique space requirement. Some SPUs may require only a few square feet, while others may require tens or hundreds of thousands of square feet.

The nature of space and the calculations required changes with each planning level. At the higher levels, space is "elastic," and the calculations may not need to be as accurate. At the lower levels, space can be more rigid but also less definite. For example, a particular machine or desk requires a certain amount of space, and the designer cannot make it fit in less space. In other instances, a piece of equipment may require a certain type of space because it has a peculiar shape, such as a U. But, under certain conditions, other items may also fit in that U shape.

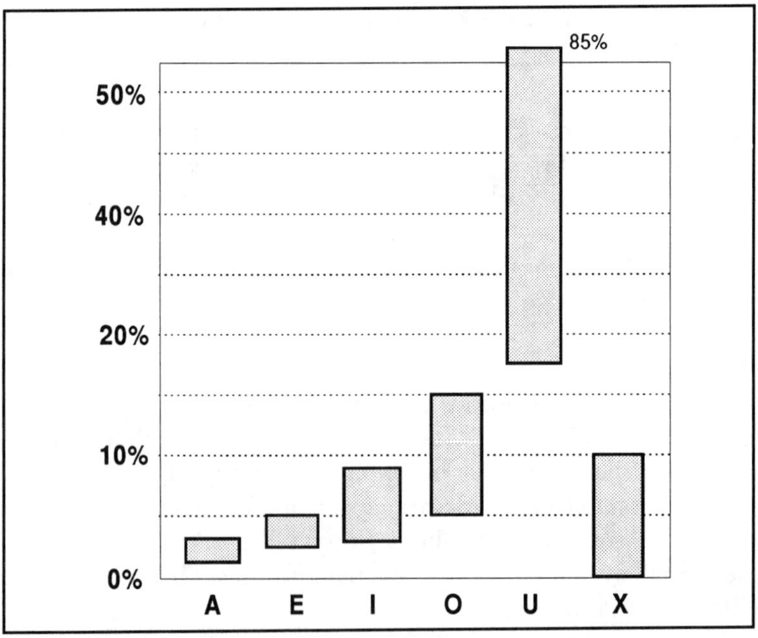

Figure 2.10 - Usual Affinity Distribution

The space plan primitive

When space is added to the affinity diagram, it distorts the diagram into the space plan primitive. It is an idealized representation and does not include design constraints.

Constraints

Design constraints are those conditions that limit an ideal space plan. Such constraints might be building size and shape, columns, floor loading, utility configurations, external features, and many others.

Space plan

The fusion of a space plan primitive and constraints produces a space plan. Several viable space plans should emerge. A set of cells, affinities, and constraints may give rise to several equally valid configuration diagrams and primitives. Each of these primitives may result in multiple macro-space plans. The nature of the design problem precludes an optimal space plan, except in the simplest situations.

The designer's experience is a key factor, for it helps him or her decide which configurations have the most potential. It helps scale the myriad of possible space plans down to a reasonable number.

Figure 2.11 illustrates the complete progression from fundamental

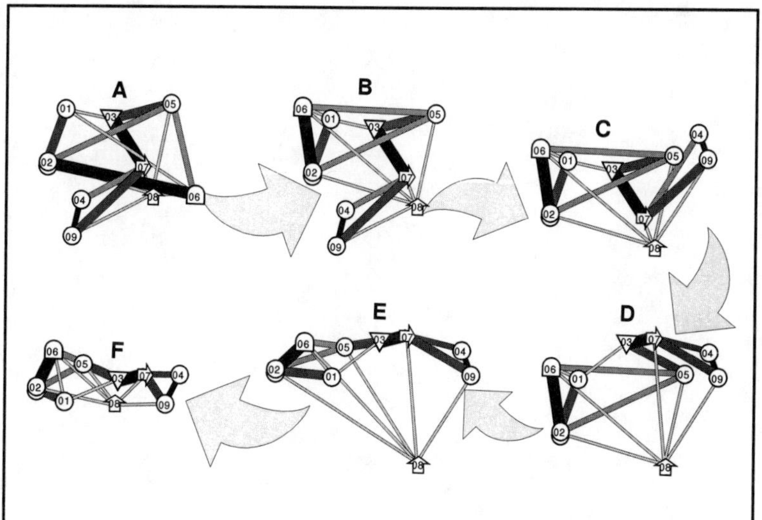

Figure 2.11 - Optimizing a Configuration Diagram

elements of cells, space, affinities, and constraints to the macro-space plan. These elements and the progression are valid for any size facility and at any level.

The design project

The elements of facility space plans are simple; execution of the tasks required to develop them is not. Rarely do the tasks neatly correspond to the development as described above. At each level of design, the approach changes to accommodate the amount of detail, available information, and the dominant issues.

At each level, an approach that fits a wide range of projects and situations can be developed. These are called *model projects*. With minor variations, the model project for a macro-space plan, for example, applies to almost any macro-space plan regardless of size, complexity, or industry. Similarly, the model projects for cell design and site planning apply to almost any cell design or site-planning project. The scope, resources, methods, formality, and time required vary according to size and complexity. The sequence, procedures, and deliverables are essentially constant. Model projects for each level of design can be found in Chapters 3, 4, 5, 6, and 7.

Chapter 3

The Macro-Space-Plan

The macro-space-plan often is the most important level of facility planning. It sets the fundamental organization of the factory and patterns of material flow with long-term effects. From personnel turnover to quality to delivery, the macro-space-plan influences almost every measure of facility and organization performance.

Done well, it is a platform for reengineering business. It can force reexamination of markets, products, and processes. It can achieve quantum improvements in productivity and profit. It can position a firm for profitability and growth. Done superficially, it can leave real issues unquestioned.

This chapter explains how to design macro-space-plans using a structured, step-by-step approach that results in a near-optimal space plan and wide acceptance of the results. This approach has several parts: *conceptual framework*, *model project plans*, *task procedures*, *conventions*, and *design tools and aids*.

Chapter 2 introduced the conceptual framework with its levels of detail that narrow the project to a manageable level. These fundamental and derived elements show how a space plan develops. Arranging the levels in phases helps plan the project. In the pages that follow, macro-space-planning—one of the more important phases—is examined. With a model project plan, tasks are arranged. Procedure diagrams illustrate how to conduct each task. The technical tools and other aids provide the means to complete each task. Figure 3.1 is the model project plan for a macro-layout. It shows the required tasks and their sequence. This model evolved from the systematic layout planning (SLP) approach developed by Richard Muther almost thirty years ago. It has been used for hundreds of projects and suffices for almost any size and type of macro-space-plan. From project to project, the depth of analysis changes along with the methods for each task, the resources, and the time. Occasionally, a project requires a few additional tasks. However, the basic structure and sequence remain the same.

Each task has a two-part identification number. The two digits before the decimal show the task level. The digits following the decimal identify the specific task, roughly in sequence. Task 03.04, for example, is the fourth task at Level 3, the macro-space-plan.

The tasks of the model project occur in three distinct groups: data acquisition, strategy development, and layout planning. These groups are near the top of figure 3.1. Two tasks, 03.01 and 03.21, are outside these groups. Task 03.01 starts the project, with plans for activities, timing, and resources. Task 03.21 is the actual selection of the preferred layout option. It closes the project and allows preparation for Level 4, the micro-space-plan.

A procedure diagram is provided for some tasks. For example, figure 3.3 is the procedure diagram for Task 03.02. Such diagrams illustrate the logic flow and sub-tasks required. These procedures are sometimes iterative. Most early layout models emphasized the third task group, where geometric arrangement takes place. Of course, this is important, but far more important is the determination of what spaces to arrange. The definition of these layout cells establishes the organization of a facility's work. Embodied in cell definition, it has far more impact on facility performance.

Figure 3.1 also guides designers through their first layouts using the system described in this chapter. The design task at hand should always be the central focus and any temptations to jump ahead prematurely to other tasks should be resisted. Completed tasks also should not be revisited. Figure 3.1 helps designers concentrate on the current task, its procedure diagram, and specific discussions. If each task is done in proper sequence, the space plan will take shape and the project objective will be reached.

It is vital to keep managers throughout the organization informed during the entire planning process, a responsibility best suited to the designer. Many facility projects result in fundamental changes and restructuring. Managers and others need time to learn new information and form new views. If they are not kept informed and involved in the learning and reasoning process, agreement and consensus will not be achieved. This could result in the rejection of an excellent layout.

There are several formal and informal ways of involving managers. Formal methods include using a steering committee to oversee and review progress and adding update meetings to the model project. A kickoff meeting can follow Task 03.01. During such a meeting, key members of the organization could review tasks and confirm that resources are available. An additional meeting, at which time factual data would be presented in a non-threatening manner, might follow the

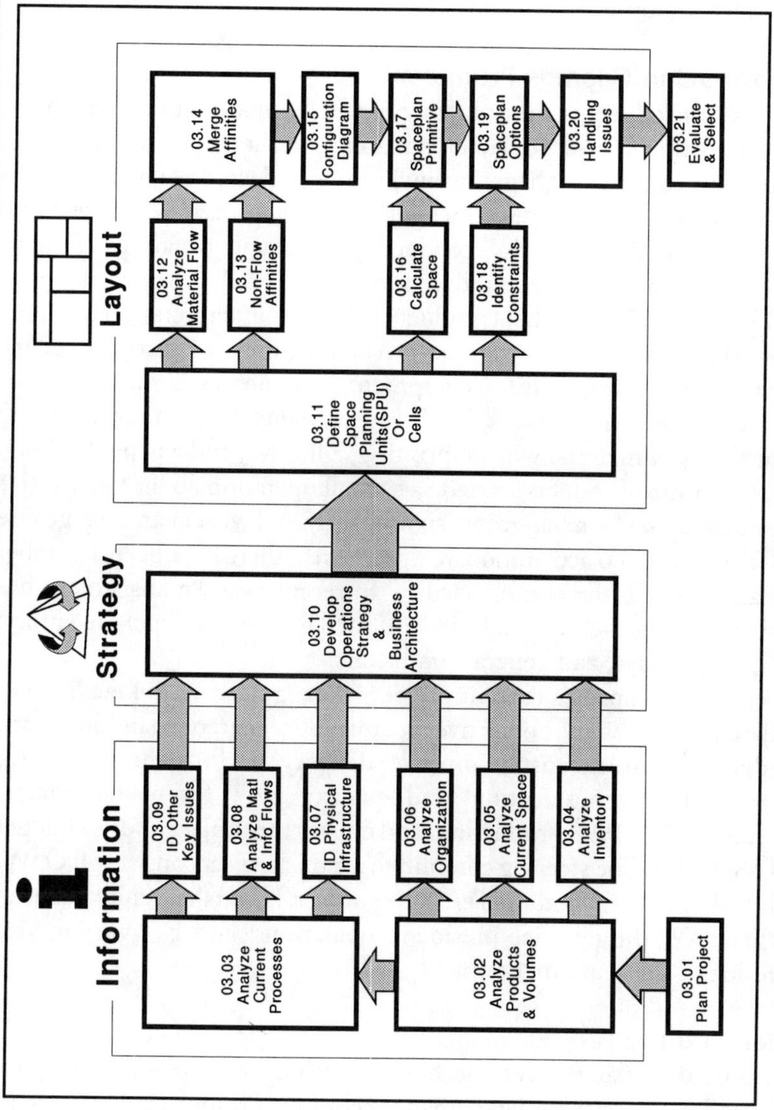

Figure 3.1 - Macro-Space-plan Model Project

data acquisition tasks.

A formal meeting is valuable for developing operations strategy and can be an important consensus builder. Agreement and commitment to the operations strategy are vital for later agreement on a facility plan. Task 03.21, the evaluation and selection of space plan options, is also a good consensus builder. Extensive interviews and informal conversations with managers and others throughout the organization are also important.

Introducing Cosmos Products

Cosmos Products is the company used as a model in this chapter to illustrate the processes of facilities planning at the macro-level. Cosmos Products converts high-grade vinyl film into decorative material. The firm has two broad product lines with different processes, markets, and distribution channels. Roll products—pin-striping material in many colors, patterns, widths, and combinations—sell in the automotive aftermarket. Custom sheet products sell to manufacturers that use them for labels, logos, and decoration. Manufacturers of campers, boats, chain saws, and agricultural equipment are typical customers. Cosmos often prepares the artwork for these customers. Custom products are flat sheets of material with imprinting, adhesive, and a paper backing.

Cosmos Products started as a small operation about twenty-five years ago. The firm has grown significantly each year at an average rate of 22 percent. To accommodate this growth, there have been a number of additions to the current facility. In recent years, management has experienced difficulty that has manifested itself in too much inventory, shipping delays, and general confusion.

The company's current project is reengineering the facility and related processes. The objectives are to: reduce material handling costs; reduce operating costs; improve delivery performance; improve teamwork, communication, and quality; allow for new products; accommodate 1998 production; and deliver the project under a budget of $800,000. The steering committee for this project consists of: O. W. Holmes, president and chief executive officer; J. Marshall, chief financial officer; W. Burger, vice president, operations; and E. Warren, vice president, sales and marketing.

Planning the project

Task 03.01, "Plan Project," develops a specific project plan. Developing a sound macro-space-plan demands significant resources. In this step, the disposition of those resources is mapped out. The model project in figure 3.1 works for almost every macro-space-plan project, whether large or small. Small or simple projects may need less formality, rigor, and documentation than larger or complex projects. Nevertheless, the essentials of each task must be done.

Step 1 establishes the key decision-makers for the project. After they are interviewed and their objectives are established, the time and resources needed for each task should be defined. With the above information and the model, the designer then plans the project. Project planning software is useful for this task, although for most macro-space-plans, a simple Gantt chart will suffice. Figure 3.2 is the schedule for a new macro-space-plan for Cosmos Products.

In addition to statements outlining tasks, elapsed time, and responsibility, the deliverables for each task should be identified. A deliverable is a tangible output for the task. A written summary of findings is a valid deliverable, as are a material flow diagram and physical infrastructure checklist. "Understand material flow" is not a valid deliverable because there is no way to see, measure, or judge completion. The designer should confirm that these deliverables accurately reflect the intentions of key decision makers before proceeding.

For Task 03.01, "Plan Project," the deliverables are a task list, a Gantt chart, and a summary that includes the project objectives. A PERT chart is useful but not necessary.

The typical time frame for completing a project of Cosmos's size and complexity is about forty working days. Almost half is used for information acquisition and strategy. This provides a firm foundation for the layouts to follow, thereby eventually reducing total project time. This is sometimes difficult for impatient managers to accept because it takes longer for a space plan to appear. However, far less time is spent on changes and debate. Moreover, a consensus for the plan is more likely when all participants have been through the information and strategy stages.

Information acquisition tasks

Once there is approval for the project plan, the first set of tasks involves the gathering of information, both quantitative and qualitative, needed to develop sound macro-space-plans.

This phase has another purpose that may be more profound and less obvious: raising awareness throughout the organization and asking difficult questions that many in the organization may not have considered previously. At this time, the process of building support and consensus for the outcome is begun.

In addition to analysis skills, space plan designers must have an understanding of individual and organizational psychology. Consensus

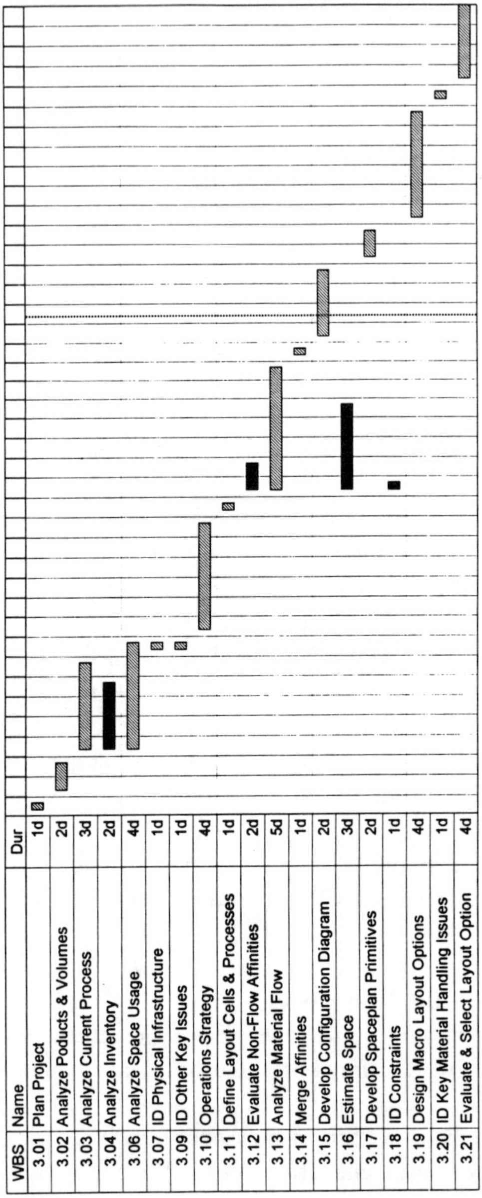

Figure 3.2 - Facility Re-Engineering

means time and common understandings that must begin early in order to bear fruit at the end of the project.

Product-volume analysis

Product-volume (P-V) analysis examines the current and future time frames for the products and their volumes. This analysis helps the designer understand the relationships between various products. High-volume and low-volume products, for example, may require different equipment and production modes. The analysis also defines future requirements, helps select the best planning horizon, and allows for changes beyond the immediate space plan.

The results of the P-V analysis provide important input for many later tasks, and, therefore, should be completed early in the project. Facility designers that have been long-time employees sometimes believe they know the products well enough to skip this task, but this is not recommended.

The procedure diagram for product-volume analysis is in figure 3.3. Block 1 documents the gathering of information. This may be accomplished in the following ways: visually examining a range of finished products; reviewing sales catalogs and other information for an overview of the product line; and interviewing sales and marketing people. It is also important to obtain overall sales volume history (usually, five to ten years is adequate). Where markets and technologies are changing rapidly, two to three years may be a more appropriate time frame.

Sales forecasts for the following five to ten years should also be requested. An absence of this information indicates uncertainty. It may require multiple contingencies in the facility plan. Unfortunately, sales people and other managers may be unwilling to commit to a forecast. In such a situation, high, low, and optimistic forecasts could be asked for, with the explanation that they are needed for facility planning purposes and extreme accuracy is unnecessary.

A request for a sales forecast may touch off a flurry of executive activity because the requested information may not exist or may be questionable. Generating the numbers will help build management awareness. It is sometimes the beginning of an important strategic debate that ultimately leads to better facility plans. This debate also can lead to important and profound changes in management thinking.

In Block 2, the forecast data is plotted on a line chart along with sales histories. If they are available, optimistic and pessimistic forecasts should also be added. After examining the chart, plotting a regression line like that in figure 3.4 may be helpful. Where seasonality is a concern, a separate chart could be used to show monthly sales for the past two to four years. Visual presentation is more meaningful than a list of numbers. A simple chart often reveals previously unrecognized trends.

In Block 3, the products are examined for appropriate grouping. If the facility will only produce a few products, such grouping is unnecessary. Most facilities, however, have many products or variations in anywhere from three to fifty groups. Preferably, these groups have common manufacturing characteristics as well as customer requirements. Sometimes the distribution channel determines sales groups.

During this task, the groups may have either a marketing or manufacturing orientation, or both. A marketing orientation means the items within a group are similar for the customer. A manufacturing orientation means the items within a group are similar for manufacturing purposes. These groups may or may not be the same. Sometimes

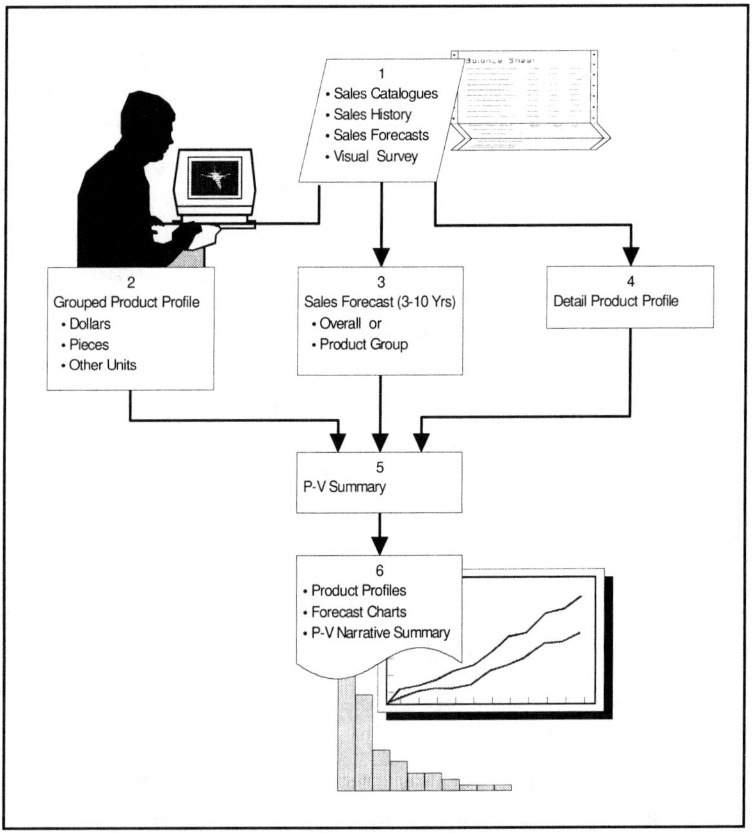

Figure 3.3 - Task 3.01, Product-Volume Analysis

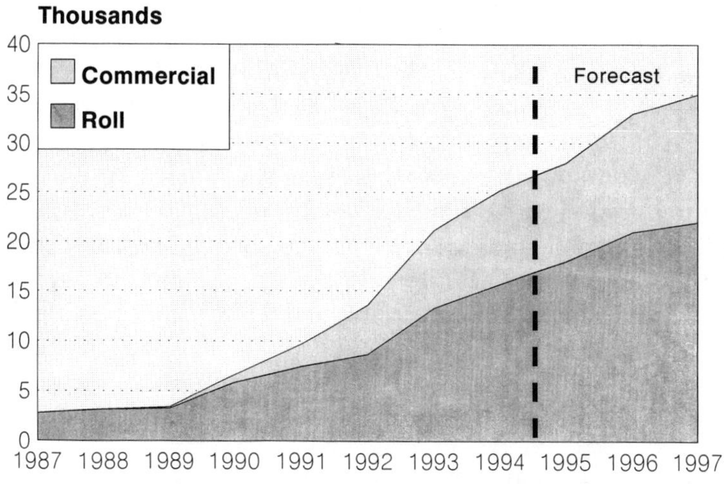

Figure 3.4 - Sales History/Forecast

operations people adopt product groups originally devised by marketing. This can complicate the manufacturing process unnecessarily.

Once the groups have been determined, a grouped product profile should be prepared. Such a profile takes the form of a ranked bar chart showing sales volume for each group (fig. 3.5). Sales volume is measurable in dollars, pieces, or other convenient units. Several profiles showing different units such as tons or pallets may be helpful. A second y-axis on the chart shows cumulative percentage.

A more detailed product profile, Block 4, also might prove useful. There are situations when a product group has significant sales volume, but individual products in the group have few (or no) sales.

The forecasts and P-V analysis become the agreed upon basis for process design, space requirements, storage requirements, and material flow analysis. It is important to confirm the forecasts and other P-V data with managers and especially with the key decision-makers.

The P-V analysis can assist with the development of the manufacturing strategy. High volume and low variety suggest high-speed production line equipment. Low volume with high variety suggests a functional layout. High variety and a wide range of volumes suggest cellular manufacturing. Seasonal variation necessitates specific strategies for inventory and capacity. The section on manufacturing strategy will explore these issues more fully.

A few short paragraphs or bullets can summarize the findings from the P-V analysis as shown in figure 3.3, Block 5.

Figures 3.3 through 3.5 illustrate deliverables for the Cosmos Product Volume Task. The following is its P-V summary, another deliverable.

Cosmos Products: Product-volume summary

The 22 percent growth rate is expected to slow somewhat during the next three years. The 1997 forecast volume of 35,000 units will be the first facility planning horizon. Cosmos has about 10,000 line items in the product database. These are in 192 groups according to significant features such as base material, color, and width. Thirty-four groups represent 80 percent of sales. Of the 192 groups, 63 generate less than $200 per month of income. We may have significant opportunity to rationalize the product offering or modify our inventory policy.

Existing process analysis

Task 03.03, "Existing Process Analysis," involves tracking work product activity, or the sequences in which outside entities act on an organization's work product. For manufacturing space plans, the work product is usually a physical product. In other space plans, the work product may be intangible such as an information packet. In a hospital, the work product may be a patient.

Existing process analysis documents the process currently in place. However, if the product is new, such a process may not exist, and a similar

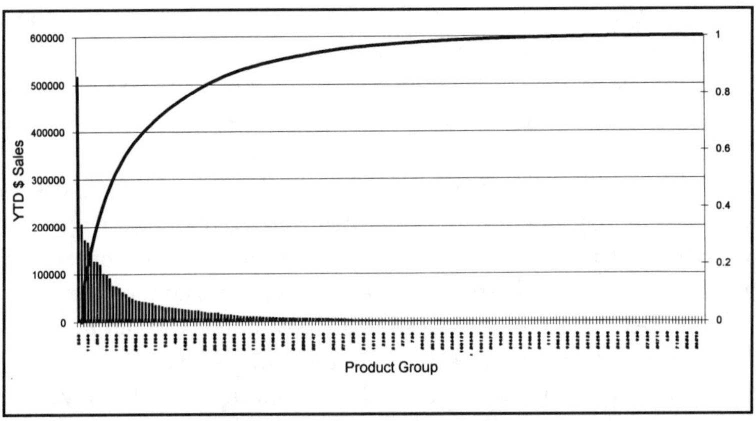

Figure 3.5 - Product Profile

product and process should be studied. If both product and process have no current benchmarks, an initial proposal for the process should be selected. The early completion of this task creates a reference point for process improvements and a space plan. The space plan designer usually performs this task with assistance from production people. This brings detailed knowledge of actual floor operations to the process.

Figure 3.6 shows the procedure for analyzing the existing process. One or more flow process charts are constructed during this task. Modified ANSI conventions (see fig. 3.7) are used in this charting system, whereby symbols represent different types of events that involve a work product.

The *operation symbol* modifies the work product in a way that advances it towards a finished state. The *transport symbol* shows a physical movement of the work product—usually a significant distance, such as ten feet or more. The *handling symbol* represents sorting, positioning, or some other short movement. Inspection checks for quality. The *delay symbol* represents something that halts the process for a time. Often, this is a work-in-process staging.

Storage is a longer wait, usually in a designated area where the location and material have records. A short horizontal line at the beginning of the process shows items from suppliers outside the process under study. Vertical lines on the chart show the sequence of events. Horizontal arrows show where several items of work product merge. Text to the right of each symbol describes the event. These notes also might indicate time, the number of people, or other relevant information.

Process charts and material flow charts should not be confused. There is a notable difference. With process charts, the symbols are not locations or workstations or even machines. Only the text has who, what, and where information. The lines do not represent movement of the work product; instead, they represent only a sequence of events.

Constructing the chart(s) means gathering initial information beforehand; this is illustrated by Block 1 in figure 3.6. Some of this comes from the P-V analysis (Task 03.02) and some from looking at the process and talking with knowledgeable people.

Block 2 (fig. 3.6) of the procedure begins addressing the question of how many and which products to analyze. It asks if there are product groups with similar processes. The answer should be based on the observations and knowledge currently available. Some situations may present thousands or tens of thousands of products.

A definite answer may not be possible without extensive analysis, which is unnecessary at this point. Suppose, for example, an injection planning facility were being planned. The plant supplies 67 molded items but each item comes in any of 19 colors. This gives a total of 1,273 item, or SKU, numbers. However, the plant uses quick color change equipment and has honed their skills in color changes. For manufacturing purposes, color is not a differentiator. The molders can make any given piece in any color or a succession of colors without difficulty. The 19 colors of each part would therefore be grouped as if they were a single product. If such groups cannot be identified, Block 3 is the next step.

Block 3 asks if there are fewer than 25 products. If there are, each

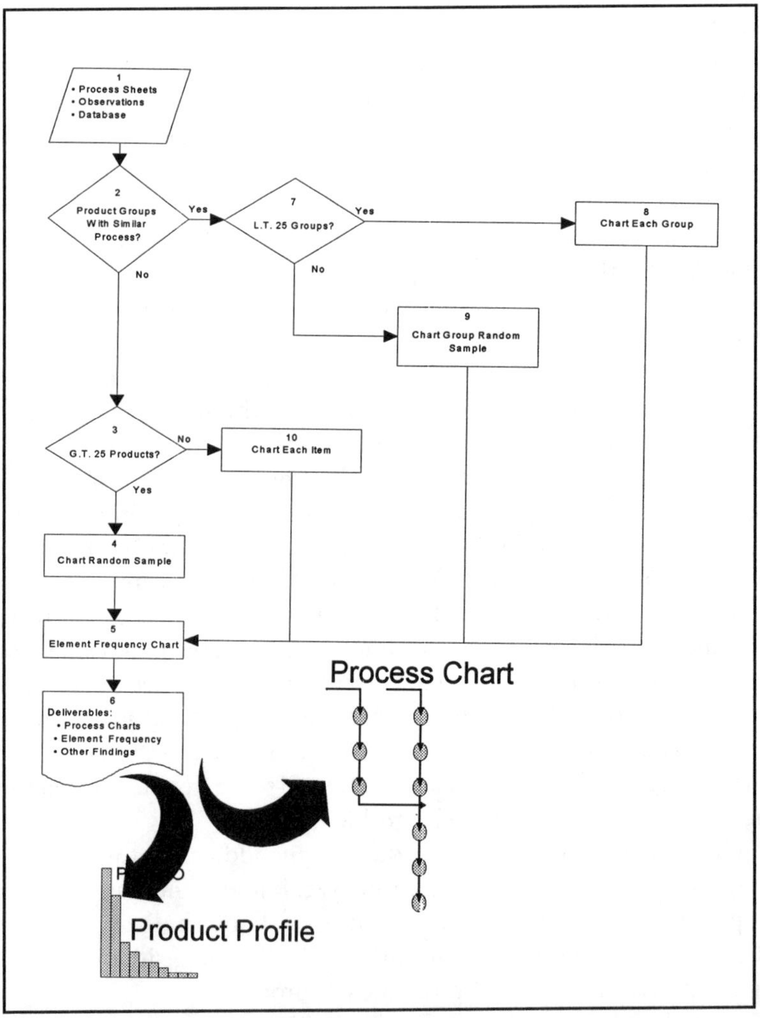

Figure 3.6 - Task 3.03, Analyze Current Process

item is charted. For more than 25, a charting sample of 5 to 25 items should be selected. This is Block 4 or Block 10. Product groups identified in Block 2 are treated similarly, resulting in Block 7. Simply substitute groups for individual products in the process described.

There are several methods of preparing the charts. For a simple process, personal observation is enough. If computerized routings are available, they may be used. A personal interview with someone who knows the process well is sometimes satisfactory. Usually, however, a group approach should be used. It captures a wide range of opinion and knowledge and helps build consensus for the chart as well as for the later space plans.

The group approach gathers the most knowledgeable people available. Together, they construct a chart that follows the material or item and records events that affect it. People often have difficulty distinguishing the product or item, workers, and machines. To help with this, they should imagine they have become the product and have assumed its role. They should then report their experiences.

All elements should be recorded. Frequently, there is an "official" process documented on routings and a computer database. Then there is the "unofficial" process—what really happens. Unofficial elements may include set downs, queues, and repairs. The group may wish to include other information on the chart such as process time or cost. When this is complete, the group should make further comments, particularly about which process elements are troublesome. Some additional questions to bring out important process issues are:

- Which elements generate the most quality defects?
- Which elements are most difficult to set up?
- Where are the largest inventory buildups?
- Which elements have the most scheduling difficulty?
- Which elements demand the most labor?

The analyst guides the group during this task by deciding:

- the level of detail for process elements;
- the number of products to chart;
- whether and how to group products; and
- whether and how to group items that go into a product.

To tally a count for each type of element, the percentage of total elements is calculated. These could be charted on a bar or pie graph. Only the operation symbol adds value. All other elements contribute only cost or time. The percentage of value-adding elements is called the value added index (VAI). VAIs frequently are in the range of 20 to 30 percent. A well-thought-out process should have a VAI of at least 60 percent.

Next, a short summary of the results should be prepared. The flow process charts, element profiles, and written findings are deliverables.

The following is the Cosmos process summary:

Cosmos Products: Existing process summary

The value added index (VAI) for roll products is 13 percent. The VAI for commercial products is 20 percent. These are both quite low. There are a substantial number of opportunities to reduce transport, handling, and storage elements.

In roll products, the processes require special equipment. This equipment is relatively small scale. Changeover times range from five to forty-five minutes.

For commercial products, process scale is very small in the manual operations at pick-and-peel. Die-cutting operates on a medium scale. Silk-screening uses large-scale presses. We may wish to investigate smaller scale processes for silk-screening.

Slit-and-sheet operations all use a single slitter that is quite fast. Both commercial and roll products use the same material. Optimizing the use of each roll saves significant wastage. It seems to dictate continued use of a common slit-and-sheet area for all products.

The process charts for Cosmos Products are fairly simple. In addition to the modified ANSI conventions, figure 3.7 shows the process for one of Cosmos's roll products—a vinyl stock material for signs and other decoration. Figure 3.8 charts the process for a multi-color, die-cut decal, a typical product from one of Cosmos's commercial markets. These decals decorate automobiles and other outdoor equipment. This single chart represents several thousand distinct products.

With complex processes, it is often tempting to combine items, thereby reducing the complexity of the chart. Simplifying the chart, however, is not the same as simplifying the process. Much of the value of a process chart is its accurate representation of the full complexity of a process. It is an important means of building consensus and understanding for a new space plan. A readable chart on large-scale drafting paper may be necessary to convey the full scope and complexity of the process.

Inventory analysis

Task 03.04, "Inventory Analysis," is important for at least two reasons. First, inventory is usually the primary or secondary capital consumer, often vying with facilities for this dubious honor. Second, almost every difficulty, problem, or defect in the business system eventually comes to rest in inventory. Inventory thus can be an indicator of the efficacy of the business system.

The inventory analysis uses financial and warehouse data. The first step in the analysis is to prepare a chart that shows historical annual inventory turns, usually for five to ten years or even further if the information is readily available. Inventory turns are the total inventory from the firm's balance sheet divided into the total sales for the previous year. Sales information usually comes from the income statement. The industry average for the inventory turn also should be listed on the chart. The inventory turns for Cosmos Products are illustrated in figure 3.9.

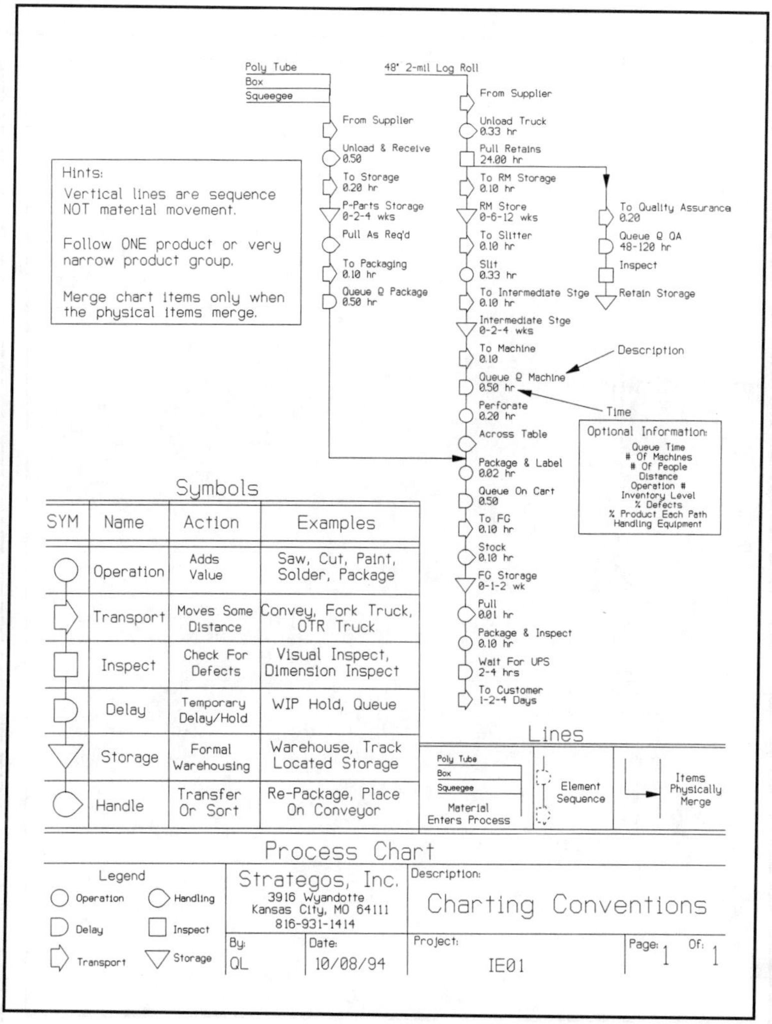

Figure 3.7 - Task 3.03, Product Process Chart

One or more inventory profiles like that in figure 3.10 should be prepared. These are pie charts or bar graphs that show the current distribution of inventory across several classifications. A production class profile should show inventory by raw material, purchased items, finished goods, and work-in-process (WIP). A product class profile shows inventory by product or product group. Other classifications, such as customer type, are useful in special situations.

What does inventory analysis determine? Trends in inventory history can help size storage areas for the new facility or layout. Such

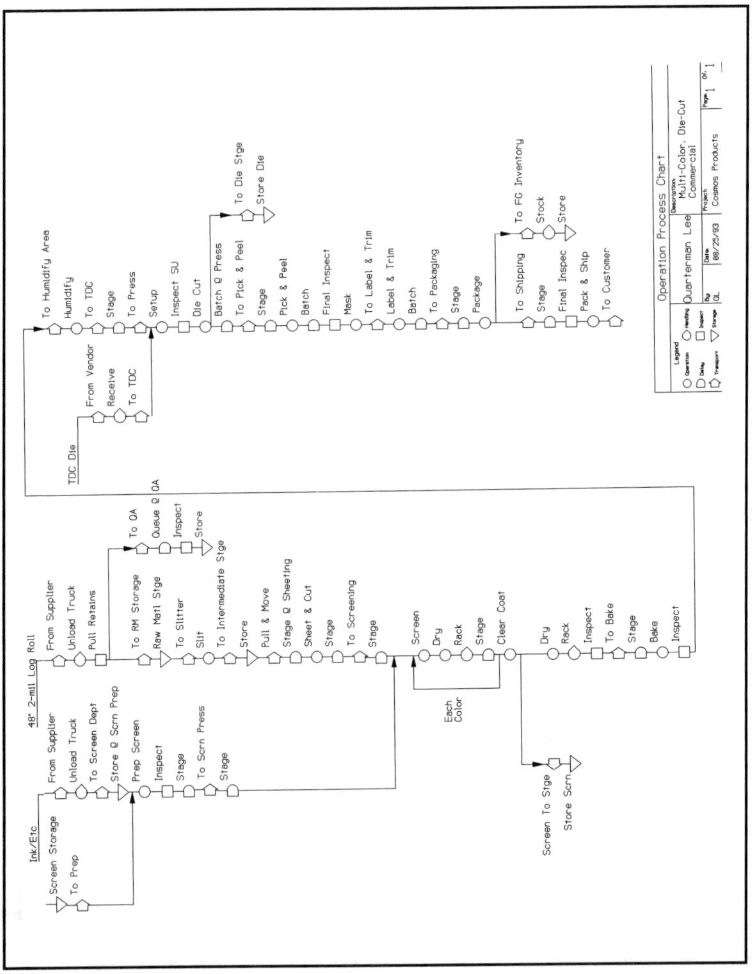

Figure 3.8 - Process Chart for a Multi-color, Die-cut Decal

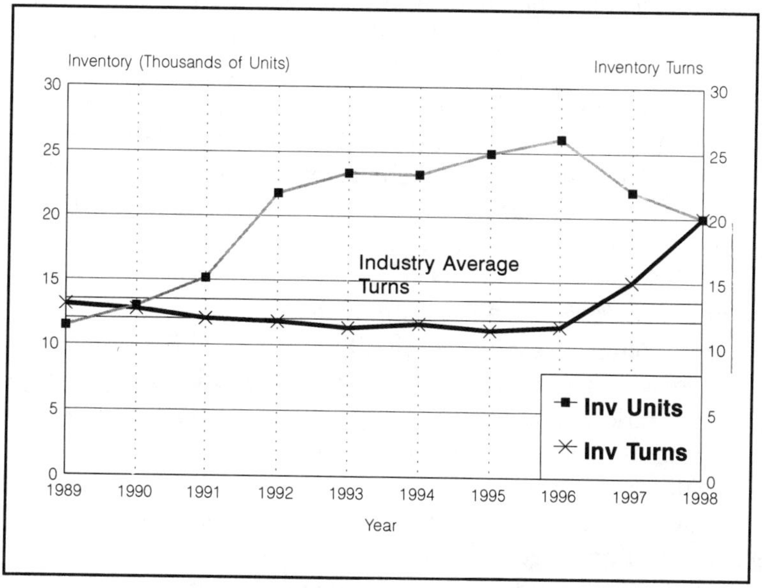

Figure 3.9 - Inventory History/Forecast

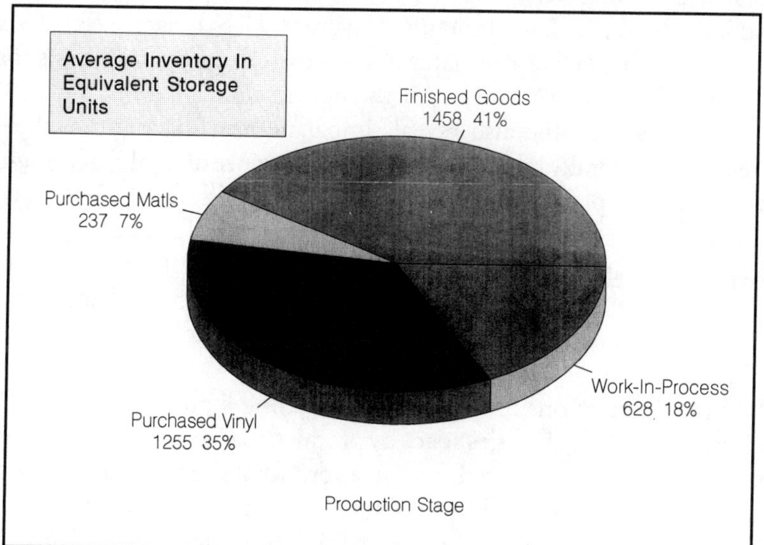

Figure 3.10 - Inventory Analysis

trends provide valuable input for the development of manufacturing strategy. An unfavorable trend might initiate a change in strategy.

The production class profile also can suggest areas for improvement. High levels of raw materials or purchased items indicate a supplier and purchasing issue. High levels of WIP indicate material movement, scheduling, or focus issues. A high volume of finished goods indicate scheduling, sales, or marketing issues. The following is an example of an inventory analysis summary:

Cosmos Products: Inventory summary

Inventory volume has increased significantly during the past six years. This increase is higher than sales growth, resulting in a gradual erosion of the turn ratio. Management anticipates that, as a result of the facility reengineering project, the number of turns will increase, and inventory levels will come down.

The inventory profile [fig. 3.10] shows the portion of inventory at each production stage. This indicates significant opportunities for reducing finished goods and purchased vinyl.

Space Analysis

The space analysis reveals current space use. The space diagrams indicate whether the existing layout is primarily functional, product-focused, or a mixture, as well as which products use line or cellular production and which use functional layout modes. This space analysis also helps define layout cells later in the project and can be a basis for space requirement calculations for the new facility.

The space profile also reveals imbalances in space use. Value-added space generally represents 60 percent or more of total space usage in the best space plans. When value-added space falls below 30 percent, there are significant opportunities for improvement. Large amounts of storage space can indicate a need for more cellular and line production, or it may show a need for scheduling system revisions. Using large amounts of space for inspection or repair may indicate significant quality issues.

When operations focus is an issue, adding a product space class diagram is useful. It classifies space by product. Each product group has a pattern or color. Space used for operations for a single product group will have only one color, while functional space used for operations for many product groups will have many colors. A product-focused layout has a “clean” product space diagram and a “messy” functional space diagram. A process focused (functional) layout has the opposite. The section on operations focus explores these issues in more detail.

The analyst usually performs Task 03.06, "Space Analysis," with assistance from those who are familiar with operations. The analysis begins with a current drawing of the facility, preferably one that shows major departments and, perhaps, details of equipment and furniture locations. The colors or patterns in figure 2.9 are then used to code marked-off space on this drawing. A typical result is the existing space diagram for Cosmos Products in figure 3.11. The area for each space class is totaled and a space class profile similar to the pie chart labeled "Existing Space Profile" in figure 3.11 is prepared.

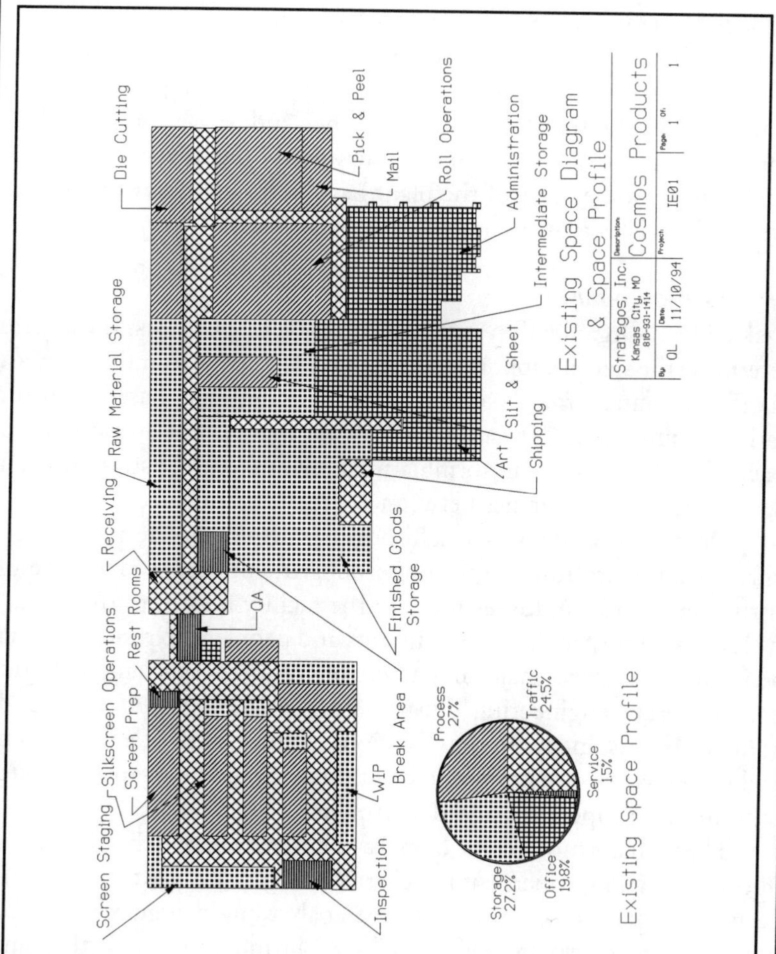

Figure 3.11 - Existing Space Diagram for Cosmos Products

The collection and presentation of this information can alert managers to key issues. As with the other information acquisition tasks, this is an important result of the space analysis. The analyst should encourage managers to begin asking such questions as:

- Why do we use 40 percent of our facility for storage, yet we constantly fall short on customer delivery?
- Why are aisles in our facility so disjointed and chaotic?
- Why does Product A require 18 percent of our facility's space but only generates 3 percent of our sales and 0.5 percent of our profit?

The following is the space analysis summary for Cosmos Products.

Cosmos Products: Existing space summary

Much of our space appears disconnected and scattered. The existing space diagram shows no clear, underlying plan. The proportions of space use are better than in many other industries, but could be improved. Significant opportunities may exist in reducing storage and traffic areas. Some parts of the plant have narrow aisles. Others have overly wide aisles that become WIP storage areas.

Organization analysis

Task 03.06, "Organization Analysis," has several purposes. It can help determine the size of support facilities such as restrooms and cafeterias. In office layouts, it may be essential for planning space based on work station requirements. It can help evaluate the current and proposed space plan. It can assist in formulating a manufacturing strategy or in identifying inconsistencies between strategy and practice.

Organization analysis usually begins with a complete and current organization chart from the personnel department. It should include all departments and employees that use the facility hdown to the lowest level. It also might include departments and people who reside outside the facility but have a major impact on operations. An example might be a corporate engineering department that designs processes and products but is in a remote location. Names and titles for each production worker are not needed, but there should be an approximate count for each supervisor and department.

These charts can become quite large and may have to be plotted on large-scale drafting paper, but the chart should not be broken into small sheets. This may be convenient for the analyst but it disguises the true nature of large, convoluted organizations. Maximum impact is the aim. Managers must develop and approve the strategic basis of the space plan, as well as the space plan itself. Figure 3.12 shows how to construct the chart.

After the organization chart is complete, the current space plan should be examined. A continuous, enclosed line on the organization chart should represent each major area on the layout, surrounding each position or department that inhabits the layout area until all positions are accounted for.

Figure 3.12 illustrates space and organization congruity. It shows consistency between the current organization and the current arrangement. People and positions in the same department generally occupy contiguous areas.

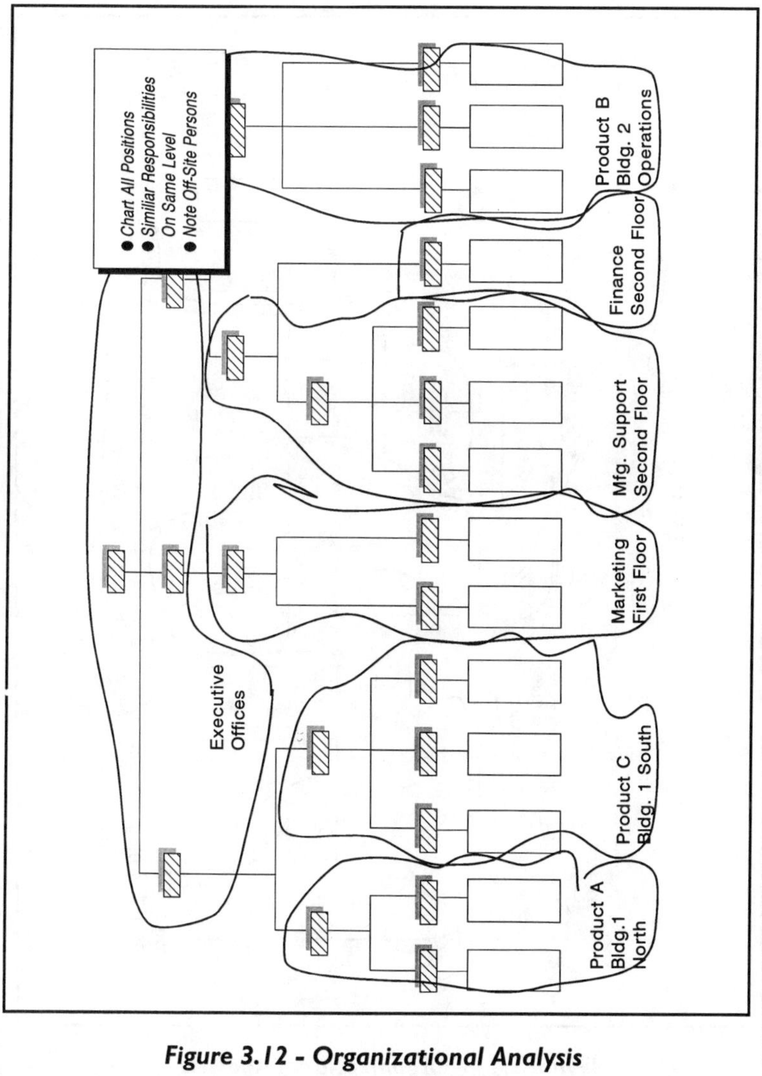

Figure 3.12 - Organizational Analysis

A messy diagram (fig. 3.13) demonstrates how many people in the same organization units are scattered through the facility. The diagram by itself does not tell us whether the facility or the organization is correct; it shows that they are inconsistent.

Identifying physical infrastructure

Physical infrastructure supports operations for all or most of the product line but does not contribute directly to the process. For this

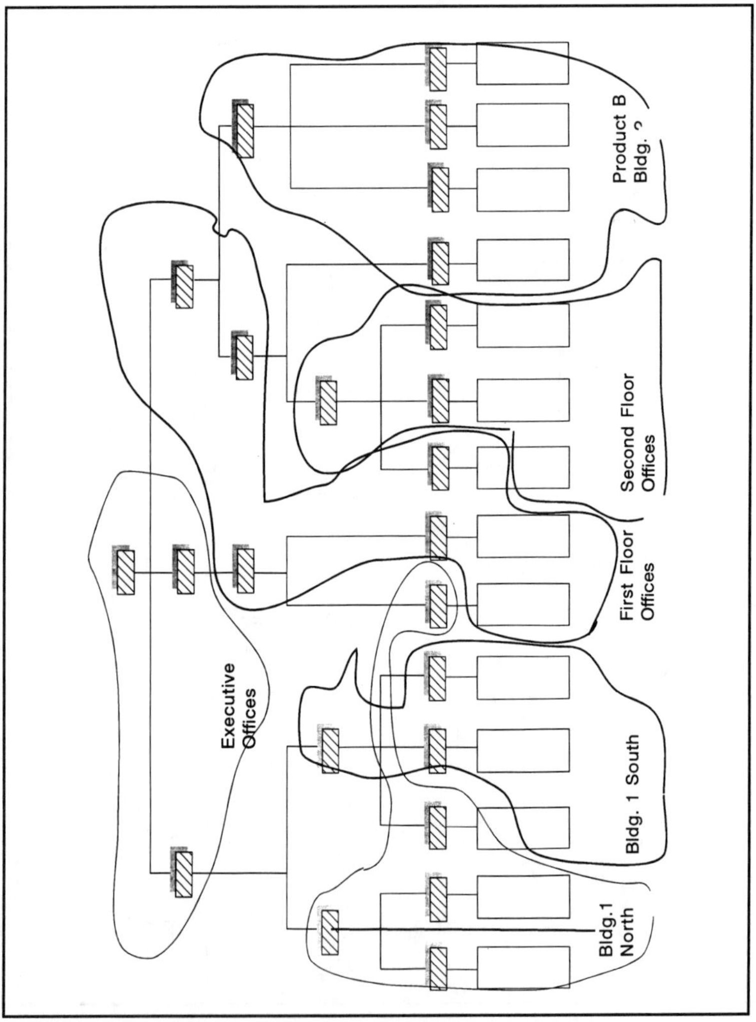

Figure 3.13 - Organizational Analysis

reason, physical infrastructure elements do not appear on the process charts. Infrastructure seldom relates to a single product or product group. Examples are: cafeteria, maintenance department, heating, ventilating and air conditioning space, and electrical switchgear rooms. These elements are necessary for operations and they are essential to the space plan, yet they are easy to overlook.

A physical infrastructure checklist (fig. 3.14) helps catalog these features. Using this form involves stepping through the list with a small group of knowledgeable people. Questions to ask are:

- Is each item in the current facility?
- Will a similar item be needed in the new facility or space plan?

This list will be input for the cell definition task later in the project.

Utilities

Electrical:
- ● Outdooor Substation
- ☐ Indoor substations(s)
- ☐ Switchgear Room
- ☐ Motor Control Center
- ☐ Uninterruptible Power
- ☐ 115v, 1-ph
- ☐ 230v, 1-ph
- ☐ 230v, 3-ph
- ☐ 460v, 3-ph
- ● 460/277v, 3-ph
- ☐ ________

Lighting:
- ● High Intensity
- ● Low Intensity WAREHOUSE
- ☐ Natural Light
- ☐ ________
- ☐ ________

Water/Sewer:
- ● Drinking Water
- ☐ Process Water
- ☐ Cooling Tower
- ☐ Water Treatment
- ☐ ________

Steam:
- ☐ Boiler Room
- ☐ Distribution System

Fuel:
- ☐ Fuel Storage
- ☐ Fuel Gas Plant
- ☐ Oil Tanks
- ☐ Coal Storage
- ☐ Other

Other Compressed Gas:
- ● Air Compressors
- ● Surge Tanks
- ● Distribution
- ☐ Oxygen Storage
- ☐ Acetylene Storage
- ☐ Other: ________
- ☐ ________

Cutting Coolant:
- ☐ Central Filtration
- ☐ Chip Separation
- ☐ Chip Conveyors

HVAC:
- ● Space Heating
- ☐ Infra-Red Heating
- ☐ Air Conditioning
- ☐ Dust Collection
- ☐ Local Exhaust
- ● General Ventilation

Other Utilities:
- ☐ Vacuum
- ☐ Other: ________
- ☐ ________

Human Resources

Administration:
- ● Personnel Office } COMBINE
- ● Employment Office
- ☐ Credit Union
- ☐ Library
- ● Training Facilities

Food Service:
- ☐ Cafeteria
- ● Vending Area
- ● Lounge/Break Room
- ☐ Executive Dining Room

Health:
- ☐ Excercise Room
- ☐ Jogging Track

Safety:
- ● Medical Center
- ☐ Disaster Alarm
- ● Fire Egress
- ● Sprinkler Controls
- ☐ Fire Station
- ☐ Tornado/Fallout Shelter

Personnel
- ● Toilets
- ● Washrooms/Showers
- ● Locker Room
- ● Water Fountains
- ● Public Telephones
- ● Employee Parking
- ☐ ________

Administration

- ● Visitor/Reception
- ● Conference Rooms
- ☐ Customer Service
- ☐ Display Room
- ☐ Training Center
- ☐ Photo/Artwork

Storage:
- ☐ Archives
- ☐ Secure Storage
- ☐ Catalogue/Display
- ☐ Microfilm
- ☐ Surplus Furniture & Equipment
- ☐ ________
- ☐ ________

Quality

- ● Quality Laboratories
- ☐ Gage Calibration
- ☐ Formulation Labs
- ☐ Discrepant Material
- ☐ ________
- ☐ ________
- ☐ ________

Internal Transport

Aisles:
- ● Pedestrian
- ● Pallet Truck
- ● Fork Truck/Tractor

Handling Support:
- ● Battery Charging
- ● Maintenance
- ☐ Central Dispatch

Built-In Handling:
- ☐ Overhead Cranes
- ☐ Freight Elevators
- ☐ Scale
- ☐ Other: ________
- ☐ ________
- ☐ ________
- ☐ ________

Production

- ● Production Offices
- ☐ Tool Room
- ☐ Tool Crib
- ☐ Clean Room
- ● Central Maintenance

Storage:
- ☐ Packaging Materials
- ☐ Lubricant/Cutting Oil
- ● Discrepant Material
- ☐ Incidentals
- ☐ ________
- ☐ ________
- ☐ ________

External Transport

- ☐ Rail Siding
- ☐ Rail Dock
- ☐ Truck Turnaround
- ● Truck Dock
- ☐ Truck Parking
- ☐ Truck Scale
- ☐ Truck Maintenance
- ● Truck Access/Egress
- ● Auto Access/Egress
- ● Auto Parking
- ☐ ________
- ☐ ________
- ☐ ________

Product Development

- ☐ Prototype Shop
- ☐ Engineering Lab
- ☐ Special Test Labs
- ☐ ________
- ☐ ________
- ☐ ________

Figure 3.14 - Physical Infrastructure Checklist

Analyzing material flow

In this task, information from the process analysis is superimposed on the current space plan. The resulting diagrams bring attention to material movement opportunities. They also indicate the need for more cellular or line production modes. This task also provides a baseline for measuring handling improvement due to the new facility or space plan.

For this task, the process charts and layouts developed in Task 03.03 and Task 03.06 should be used to select one or more items to represent typical products or parts. Lines and arrows should trace movement across the layout. The number of moves for each item should be counted and the movement distance for each item totaled. If the analysis is performed for many items, the results should be averaged. Figure 3.15 is an example. It shows long moves, crossovers, and backtracking. This facility has significant improvement opportunity. Moves between organizational departments also indicate improvement opportunity.

Managers are often unaware of the severity of material flow problems. This analysis will document these issues in a dramatic way. It helps management take another step towards consensus, understanding, and support. Other types of material flow diagrams are also useful for a more complete picture of the current material flow.

Other issues

Other issues can affect the layout. They usually arise in initial discussion or during data acquisition. Some examples are:

- a scheduling system that dictates batch movement through the plant;
- difficulties in hiring skilled people that may push a company toward automation; and
- external regulations such as those in the pharmaceutical industry that may dictate functional operations.

Experience and judgment are the best guides. At minimum, a brief listing of these issues is necessary. They may need significant analysis.

The strategic framework

An operations strategy is the dominant approach or philosophy that guides the design of the manufacturing or business system. Operations strategies often determine the competitiveness and ultimate fate of an organization. Strategy leads to structure, as well as the arrangement and interconnection of business elements. Such elements might be machines, information systems, people, or facilities.

Strategies extend over long periods—years or decades. They encompass all the products and processes, permeating every area and aspect of the organization. They affect and determine the behavior of individuals.

Operations strategy may be explicit or implicit. An explicit strategy is stated, orally or in writing. Properly promulgated, it guides decision-makers in their daily work, building a common framework for both operational and structural decisions.

An implicit strategy, by contrast, is not written or publicized as such. It often results from common understandings about what matters for the business. These understandings may be rational or senseless, effective or ineffective, consistent or contradictory.

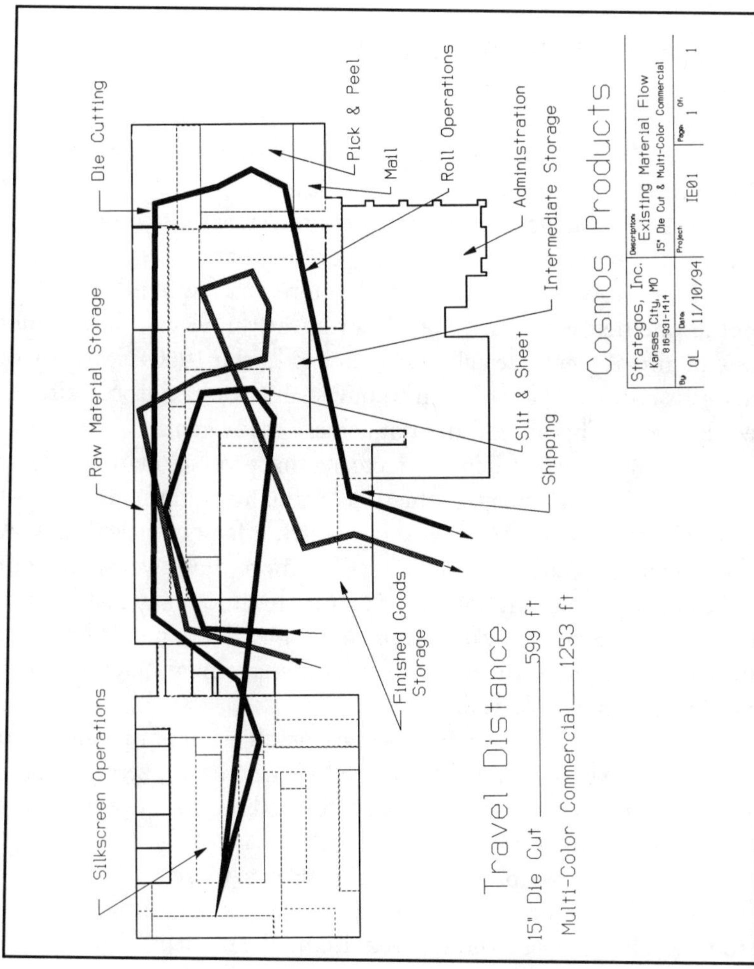

Figure 3.15 - Existing Material Flow

An implicit strategy is determined only by watching patterns of decisions and behavior over time. For example, has the organization evolved along functional rather than product lines? Is new equipment predominantly high-speed and large-scale? Is the scheduling system batch-oriented? Even the absence of patterns is a pattern.

Management is responsible for operations strategy. Top managers can abdicate the task of enunciating that strategy, but they cannot relinquish responsibility for the result.

Determining the framework of an organization's operations strategy hinges on the identification of *key manufacturing tasks*, *focus opportunities*, and *an operations strategy summary* (or statement).

Key manufacturing tasks and focus opportunities

The design of a manufacturing plant or business is like any other engineering design. It optimizes performance on some dimensions, while reducing optimization on others. The business environment and available technology place limits on the design.

This has an analogy in aircraft design. Aerospace engineers can design an aircraft that flies at Mach 3.0 or one that carries 350 people. They can design an aircraft that circles the globe on a few hundred gallons of fuel or one that lands on a 500-foot runway. However, they cannot design an aircraft that does all of the above. In the 1960s Secretary of Defense Robert McNamara tried to buy such a multi-purpose aircraft called the TFX. This aircraft did not achieve many goals.

Yet, many managers demand factories that produce many products quickly for many customers, at the highest quality and the lowest cost with output changing from day to day. Such a factory lacks focus. A business operation rarely performs well on more than two or three of these key dimensions. An unfocused factory has too many tasks or too many products or too many process technologies or too many disparate customers. It is often too large for effective management. Such a factory rarely performs any task well.

Manufacturing focus concerns the organization of products and processes. In the early 1970s, Wickham Skinner recognized that large factories with many products usually performed poorly. Several factors contribute to this effect:

1. A wider range of products usually brings more variety in the process. This requires greater complexity in handling, storage, tooling, changeovers, and skill requirements. It affects almost every facet of operations.

2. A wider range of products often must serve disparate customers and

markets. One market may regard delivery speed as a top priority while another demands quality or customization. Such varied market criteria increase the difficulty for manufacturing and decrease effectiveness.

3. Economies of scale are the usual rationale for increasing factory size. Economy of scale refers to the increasing efficiency as plants and processes grow in size and output. The idea was popularized by Henry Ford's mass production methods. Wickham Skinner coined the term "dis-economies of scale." Increasing scale brings such dis-economies as increased coordination effort, isolation of specialty departments, and isolation from customers. As a factory grows beyond 300 to 500 persons, the dis-economies of scale soon overcome the economies.

4. Larger factories have greater distances between departments. This increases material handling costs and exacerbates the isolation and coordination difficulties.

5. Unfocused factories often have extensive vertical integration. Vertical integration with a wider product range requires more disparate processes. This requires far more technical mastery than a more focused operation.

A focused factory strives for a narrower range of products, customers, or processes. The result is a factory that is smaller and has few key manufacturing tasks.

In recent years, Skinner's concept of the focused factory has been extended. Focus is an issue when organizing any combination of products, technology, and people. It applies to service operations, to factories, and to departments within the factory. It applies to workstations within each department. The issue is: by what criteria shall we divide our space, people, and machines into manageable units?

There are several possible responses. Some examples are: products, processes, markets, customers, geographic areas, and support requirements. For a more complete discussion of operations focus, refer to the first chapter of the *Handbook of Commercial and Industrial Facilities Management*.

For the macro-space-plan of a factory, the focus choice usually narrows to product or process. A product-focused plant groups operations into departments that focus on products. Each department must have all equipment and skills for all operations, yet only process a single product. This eliminates changeovers and reduces coordination and scheduling problems.

A process focus allows each department to specialize in their particular process or craft. It is a common arrangement in many plants, probably taken from the medieval craft guilds.

Many of the perceived advantages of process focus are elusive in practice, although process-focused space plans and organizations do work well in certain, specialized situations. On the whole, product-focused space plans are preferred because they have many more advantages. Designers should aim for the highest degree of product focus attainable, using process focus only when exotic skills and large scale processes make it necessary. The areas in which product focus has advantages include: cost control, coordination, material flow, management and supervision, equipment utilization, knowledge and skills, response time, flexibility, quality, and organization.

Product focus simplifies cost control because it pulls together the same or similar products and converts many indirect costs to direct. Elaborate tracking and allocation schemes are often unnecessary. Because a process-focused operation must address a wider product variety, allocation of indirect costs is more difficult.

Product focus simplifies the coordination of sequential processes. Operations are in small areas, reducing the complications of distance and isolation and simplifying personal communication between operations. Because the product range is narrow, only a small variety of problems and issues will arise. Product focus often uses simpler methods for production control such as Kanban and direct link. In conjunction with MRP-type systems, it reduces the number of work centers the system schedules.

When compared to process focus, material flow reductions of 80 to 95 percent are common for product-focused operations. There are fewer interdepartmental moves, and distances are shorter. Variable flow paths often become fixed upon conversion to product focus. This allows the use of simpler handling devices, such as conveyors, or even manual handling. Process-focused space plans often require expensive automatic guided vehicle systems or even more expensive fork trucks.

Because of the smaller product range and better communication, product focus simplifies management. Product-focused cells often require little or no management because they naturally encourage teamwork. Product focus achieves the shallower organizations now in vogue. More emphasis is placed on products and customers rather than departmental loyalties.

In theory, product focus uses more equipment for the same output than process focus. However, firms seldom realize the theoretical utilization advantage of process focus because of the complex scheduling required. In practice, there are several approaches to mitigate the apparent under-utilization of product focus. One way is to design cells that maximize the use of major equipment while sacrificing usage on less expensive peripheral equipment.

Product focus requires a wider range of employee skills and knowledge. This may place large training burdens on firms that convert from process focus. However, the teamwork and job enrichment that result reduces turnover. Process focus, on the other hand, allows concentration on process skills, and highly complex and technical processes sometimes need this concentration.

Process-focused organizations typically have very long throughput times. As a result, they cannot respond quickly to changes in product mix, volume, or special requirements. Many process-focused plants counter this with extensive inventories, even though inventory is expensive and rarely reduces the response time on customized products. Product focus allows firms to eliminate finished-goods inventory while improving delivery performance and reliability.

Process focus is more flexible, at least in theory. However, several means exist to achieve good flexibility in product-focused layouts. For example, the use of small-scale, mobile equipment can allow product focused cells to be formed, disassembled, and re-formed new products.

Product focus generally achieves high quality levels. This results from quick feedback, good communication, easy coordination, and high commitment. Process focus sometimes may have a quality advantage for complex or technical processes.

Product focus is most compatible with newer approaches based on teamwork and empowerment. Process focus lends itself to traditional command and control management styles, often requiring a substantial hierarchy to deal with increased coordination and complexity.

The concepts of focus and key manufacturing tasks are interrelated. Focus identifies the most important dimensions and optimizes them. The business addresses a narrower market, but addresses it very well. The key manufacturing tasks state what manufacturing must do well to survive in the market.

Process elements are the equipment, people, and operations that add value to a product. They directly transform materials, information, and parts.

Operations strategy summary

A sound operations strategy addresses four areas: *mission*, *process*, *infrastructure*, and *facilities* (physical infrastructure). The site mission states, in a few paragraphs, the purpose of the site. It identifies customers, products, and processes. It defines one to three key manufacturing tasks that directly correlate to success in the marketplace. The mission statement also might address important external issues such as environmental policy. The remainder of the operations strategy summary flows from the mission statement. It states how the firm intends to achieve the key manufacturing tasks.

Infrastructure supports the process but does not directly affect the product. Non-physical infrastructure covers a wide variety of elements. It refers to people and information systems. Examples are: scheduling systems, training operations, personnel departments, and tool design capability.

Physical infrastructure, is tangible and is generally synonymous with facilities. Buildings, utility systems, roads, and docks are not directly in the process stream; rather, they support all processes.

A good summary addresses each major topic at a policy level. Few companies have a strategy summary sufficient for facility design purposes. For such purposes, those elements of strategy and structure that relate to facilities need to be emphasized. For example, compensation systems have major effects on organizational behavior but little consequence for the facility plan.

Figure 3.16 provides a structure for a strategy statement. Such a statement normally consists of one to four pages summarizing the firm's intentions for each structural element. Figure 3.17 is the statement for Cosmos Products.

The absence of an effective summary presents the facilities planner with difficult options:

- proceed without a summary;
- guide management as they develop a strategy;
- write a summary based on an idea of what it should say; or
- write a summary based on what probably will happen.

Development of a strategy summary relies on all of the information collected during the first task group. Even this may be insufficient for a complete statement. Strategic development is a high-level task that encompasses almost every aspect of the business.

Identifying operations strategy

Figure 3.18 is the procedure diagram for Task 03.10, "Identify Operations Strategy." Block 1 of the diagram begins with the assembly of information in a form suitable for a report or presentation. The *operations strategy outline*, Block 2, is also needed.

Block 3, the current position summary, shows the company's

present status. A management team should assist with this step or at least concur that the summary reflects the company's current situation.

Blocks 4 through 6 determine management's readiness for change. It is not enough for management, or certain individuals, to express a need for change. The management team also must be capable of carrying through with change. The success of a space plan depends on the organization's ability to support it. For example, a cellular space plan that depends on kanban production control and small lot sizes needs

TechTools

Strategy and Structure Key Issues

1.0 Site Mission

1.1 Site Focus
- 1.2 Products
- 1.3 Markets
- Volumes
- Geography

1.4 Multi-Site Integration
1.5 Key Manufacturing Tasks
1.6 External Strategic Issues
1.7 Political
1.8 Environmental
1.9 Community Involvement

2.0 Process

2.1 Production Mode(s)
- Project
- Functional
- Cellular
- Toyota
- Line
- Continuous

2.2 Process Scale
2.3 Setup/Lot Size
2.4 Capacity
- Lead
- Track
- Lag
- Reserve

2.5 Quality Capability
2.6 Technology Level

3.0 Infrastructure

3.1 Quality Approach
- Quality Police
- Quality at Source

3.2 Personnel Policies
- Technical Skill Depth
- Technical Skill Breadth
- Interpersonal Skills
- Employment Security
- Compensation
- Training
- Performance Measurement
- Safety
- Ethics

3.3 Organization Structure
- Functional
- Product
- Other
- Depth

3.4 Organization Style
- Exploitive
- Bureaucratic
- Consultative
- Participative

3.5 Accounting Policies
- Process
- Job Costing
- Activity-Based Accounting
- Overhead Allocation
- Decision Criteria
- Knowledgebase Investments
- Inventory Accounting

3.6 Production Control
- Make-to-Orders
- Make-to-Stock
- Physical Link
- Broadcast
- Kanban
- MRP
- Re-Order Point

3.7 Supplier Policies
- Selection Criteria
- Single / Multiple Sources
- Contract Time Horizons
- Scheduling Approach
- Shipping Policies
-

4.0 Facilities

4.1 Site Focus
- Product
- Process
- Market
- Geographic
- Other

4.2 Site Location & Size
4.3 Transportation Access
4.4 Utility Systems
4.5 Expansion Policies
4.6 New Product Flexibility
4.7 New Process Flexibility
4.8 Resale / Disposal Policy
4.9 Hazardous Waste Policy
4.10 Environmental Issues

Figure 3.16 - Strategy and Structure Key Issues

rapid setup techniques and participative management. Management may not have the wherewithal to adopt these techniques.

If management has an acceptable strategy summary, this summary should be used in subsequent space plan work. If management expresses a desire for operations strategies different from current practices, their readiness to make required changes must be evaluated. This is a difficult and sensitive decision that requires years of experience in institutional change. Designers who lack such experience should seek counsel from the management team or others in the organization.

Mission

Site #1 has two distinct missions corresponding to our two major markets. These are:

Roll Products- Supply our customers with the largest variety of roll products in our industry with 24-hour shipment and quality within the top 20% in our industry.

Commercial Products- Supply high quality vinyl decals and appliques to small and medium size manufacturers. Originality and quality of artistic design is an integral part of our product mix. We expect to ship 80% of all orders within two-weeks at a reliability of 98%. Our products will conform to recognized quality standards.

At Cosmos we expect to be a good neighbor and integral part of our community. Cosmos should be known as a satisfactory employers.

Process

Roll Products- Cosmos will strive for a product-focused operation with the exception of primary slitting which serves both Roll and Commercial operations. We will have a mix of large and small scale equipment in Group Technology cells. Rapid setup is an important priority for equipment selection and operation. We will attempt an average equipment utilization between 60% and 85%. We will add process capacity 6-12 months ahead of demand. Processes should have a minimum capability index of 1.4. We will move gradually to higher technology processes provided they are consistent with our focus strategy, cost justified and have adequate support.

Commercial Products- Cosmos will strive for product-focus within the limits set by process and environmental requirements. This may dictate physical separation between silkscreening and subsequent operations. Primary slitting and sheet cutting will remain process focused. Our process scale will be a mix of large and small corresponding to the order mix. Rapid setup is an important priority on the smaller scale, low volume processes. We will attempt an average utilization of 80%-90% on large-scale silkscreen printers and 50%-70% on other equipment. We will add small-scale equipment in advance of demand and large-scale equipment when demand is proven. All equipment will have a minimum capability index of 1.4. We will strive for the latest and highest technology level on large-scale silkscreening.

Infrastructure

Cosmos will strive for a product-focused, shallow, consultative and informal organization. We will gradually move towards a participative team-based organization over the next five years.

Our accounting system should accommodate activity-based costing using cost drivers for overhead allocation. We will use project costing for commercial work and process costing for roll products. We recognize the limitations of conventional accounting systems for management decisions.

Production control will use MRP-type scheduling for commercial products and suppliers. We will use kanban systems for internal scheduling of roll operations. Commercial products will be strictly make-to-order. Roll products will use small finished goods stocks for the highest volume 80% of line items. The remaining 20% of low-volume roll products will be made to order. Unusually large orders of roll products will have extended deliveries and be made to order.

At Cosmos we will strive for long-term relationships with reliable suppliers. We will select suppliers on the basis of quality, delivery reliability and cost in that order.

Facilities

Site focus will follow corporate-level strategies as new sites develop. All sites, now and in the future will have a maximum of 200 employees. Site #1 requires only limited capability for new products and processes. Significantly different processes, such as a casting operation, should have a separate site.

Figure 3.17 - Physical Infrastructure Statement

An organization may not want change or may not be positioned for change. If so, the operations strategy summary should identify the approach that is most likely to be adopted in practice.

If the organization desires and is ready for significant change, initiating a strategic debate is a good idea. This debate should conclude with a proposed operations strategy summary that will help the space plan designer carry out the new strategy during the facilities plan.

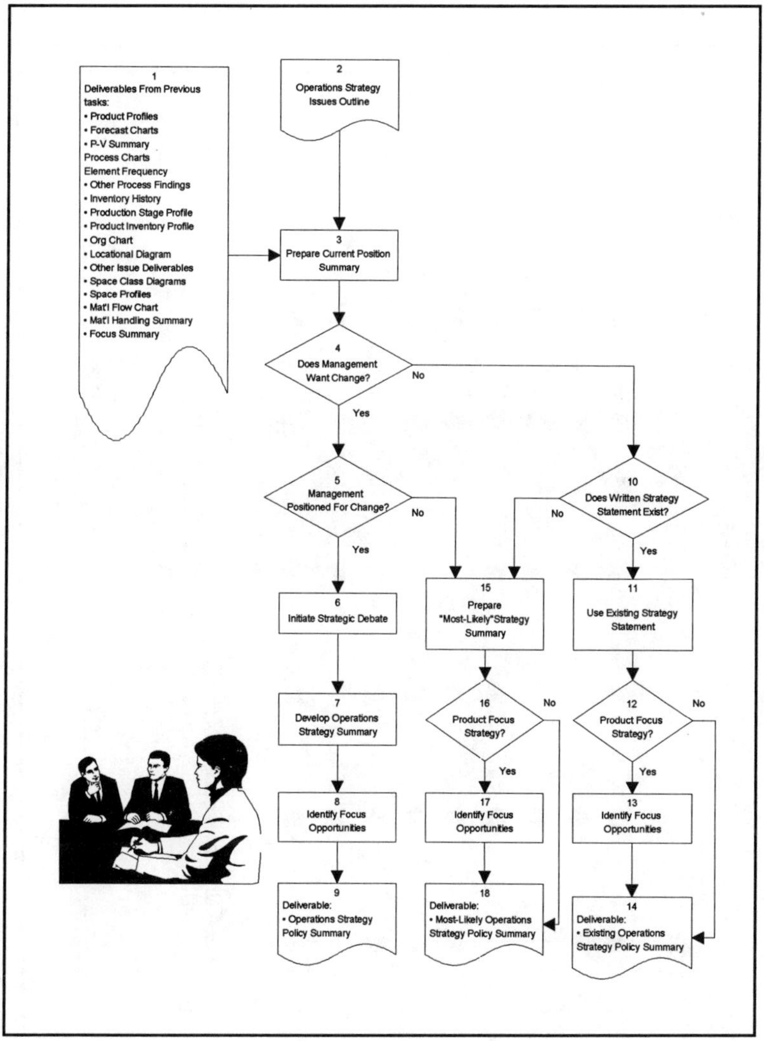

Figure 3.18 - Task 3.10, Identify Operations Strategy & Structure

One sub-task of Task 03.10 is the identification of focus opportunities. The concept of focus applies to space plans, organization structures, and other elements of the enterprise. Developing an appropriate strategy for facility planning means identifying the most appropriate focus for the facilities at each level. This is not a final analysis. Rather, it guides and gives preferred directions to space planners as they proceed with their work.

The flow process charts from Task 03.03 can help sort this out. Figure 3.19 shows the process charts for the manufacture of cosmetic

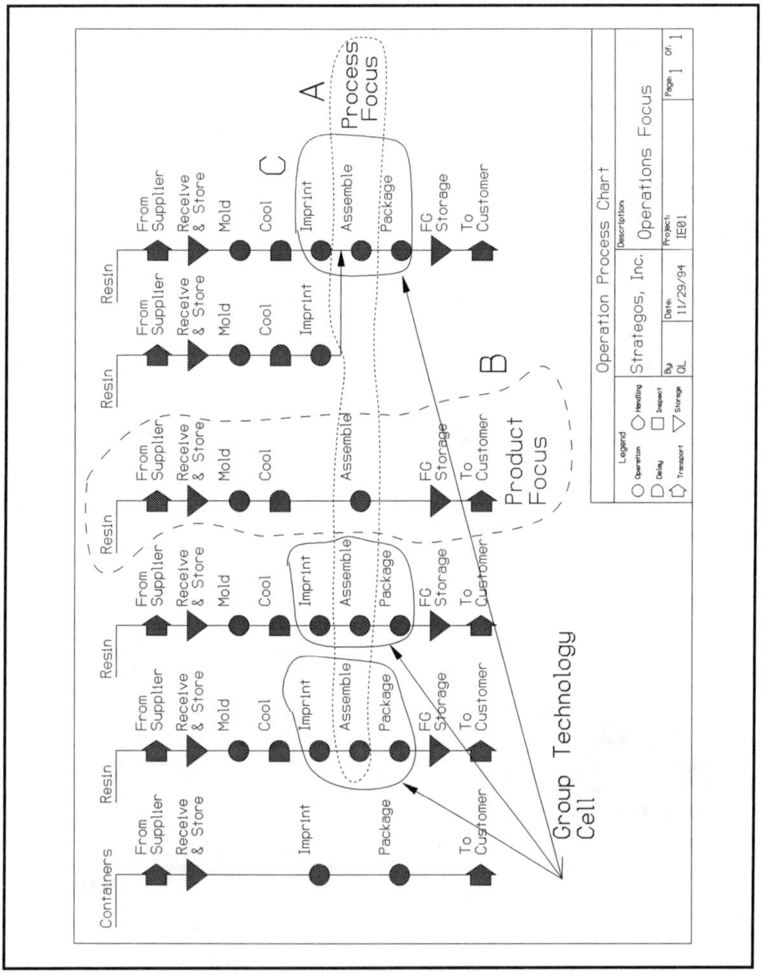

Figure 3.19 - Task 3.03, Product Process Chart

containers. The principal operations are injection molding, assembly, transfer printing, and packaging.

There are four basic containers. Each has varieties of color and print. Because color and print changeovers are fast and easy, the plant considers only container styles different products.

Process elements usually touch the product. At Cosmos Products, people assemble, lathes turn, and molding machines convert powdered resin to solid components. The selection, arrangement, and operation of these and similar elements are part of process design. A purely process focus would group the molding machines into a molding department, printing into a print department, and assembly into an assembly department. Each department would perform operations on all four products. Envelope A shows how assembly operations might fit into an assembly department.

Envelope B shows how all operations for a single product would be coordinated.

Sometimes focusing purely on product is impractical. Instead, group technology cells might be used, whereby a series of operations for several products takes place in a single cell. Envelope C shows how these operations might be coordinated.

Various mixed approaches are common. For example, process focus might be used for receiving, shipping, and molding, while other operations might have a product focus.

Figure 3.20 shows the procedures for identifying focus opportunities. The first is preparation of process charts for all products. This may be done on paper, or, for large numbers of products, forms of computer analysis can be used.

Next, products that are candidates for a plant-within-plant are identified. These would have enough volume to justify separate equipment, people, and infrastructure. Any such products should be set aside and removed from further consideration. A plant-within-plant (PWP) is a self-contained production facility within the walls of a larger facility. Ideally, a PWP is completely independent with its own supporting infrastructure.

We then search for products (or components) that have similar operation "strings"—groups of operations that can use the same equipment, the same people, and perhaps the same tooling. Is there adequate volume to justify dedicating equipment, people, space, and infrastructure to this group? If so, these products or components must be removed from further consideration and assigned to a group technology (GT) cell.

This procedure continues until the only remaining products and operations are those too small and varied for dedicated plants or GT cells. Process-focused cells are then developed for these items. An alternative is a job-shop department similar to a prototype shop.

Some space plans involve a great number of products and components, perhaps thousands or tens of thousands. In these situations, practicality may not allow a detailed analysis at this point in the macro-

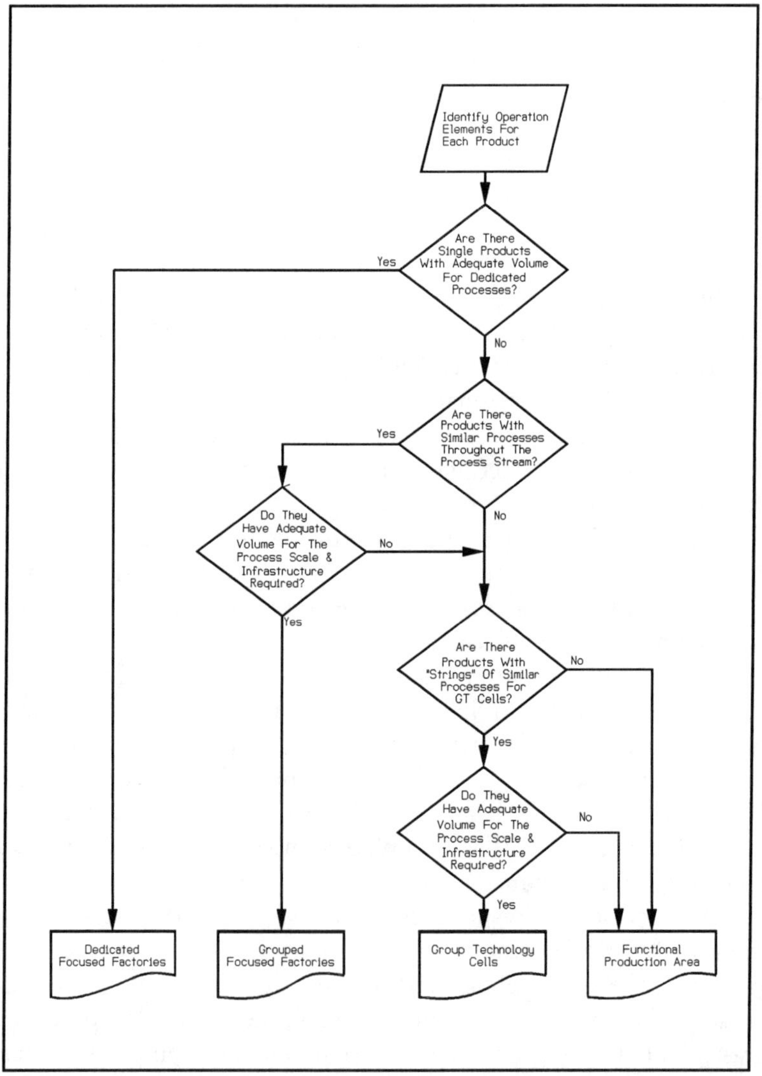

Figure 3.20 -Procedures for Identifying Focus Opportunities

space-plan project. The objective for this sub-task is to identify opportunities, not to make decisions.

For Cosmos Products, the two distinct product lines—roll products and commercial products—separated naturally into focused factories. However, the slit-and-sheet operation served both product lines. Many log rolls, when slit, became stock for both product lines. The narrow widths necessary for roll products are a natural byproduct of slitting for the wider commercial items. Maintenance, quality, and several other functions cannot be split economically. For this reason, the design team's aim became two semi-focused factories within the same facility, i.e., plants-within-plant. Several functional areas serve both focused factories.

Roll product operations lend themselves well to GT cells. However, the large number of items precludes a complete GT analysis at the macro-space-plan level. The team therefore developed a composite cell for roll operations, with the intention of analyzing the process in greater detail and designing GT sub-cells at the next design level.

In commercial products, the silk-screen operations call for tight environmental control. In addition, the existing silk-screen presses use large-scale, high-technology equipment. The team decided to put two composite cells in the commercial area. The first composite cell would include silk-screen and any related operations in the controlled environment. A second composite cell would have post-silk-screen operations such as thermal die-cut, pick-and-peel, and packaging. These smaller-scale processes would be arranged into GT cells.

Designing the space plan

With adequate information and an agreed-upon strategy, the actual space plan can be designed. The activity to this point may have consumed as much as half of the time and resources available to the project. Nevertheless, these expenditures were good investments. Managers from all areas have new perspectives. The factual data has tempered emotions. As the space plans develop, debate should be constructive and rational. The final selection will enjoy wide support thanks to management's broader understanding of both business and technical issues.

Defining space plan cells and processes

Task 03.11, or the definition of space planning units (SPUs), is the most fundamental and important task in space planning. It establishes the organization of space and must fit with a corresponding organization of people and processes. Moreover, all subsequent work flows from this task. An omission or error invalidates all of the work that follows.

A procedure chart for this task is in figure 3.21. Blocks 1 through 3 call for assembling deliverables from all previous tasks. Block 4 reviews operations strategy. If the strategy statement favors a process-focused (functional) space plan, planners should proceed to Block 5 and skip Blocks 11 through 23.

In Block 5, functional and support cells for the space plan are identified by examining the cell definition summary (fig. 3.22), the space analysis, the infrastructure checklist, the process charts, and the

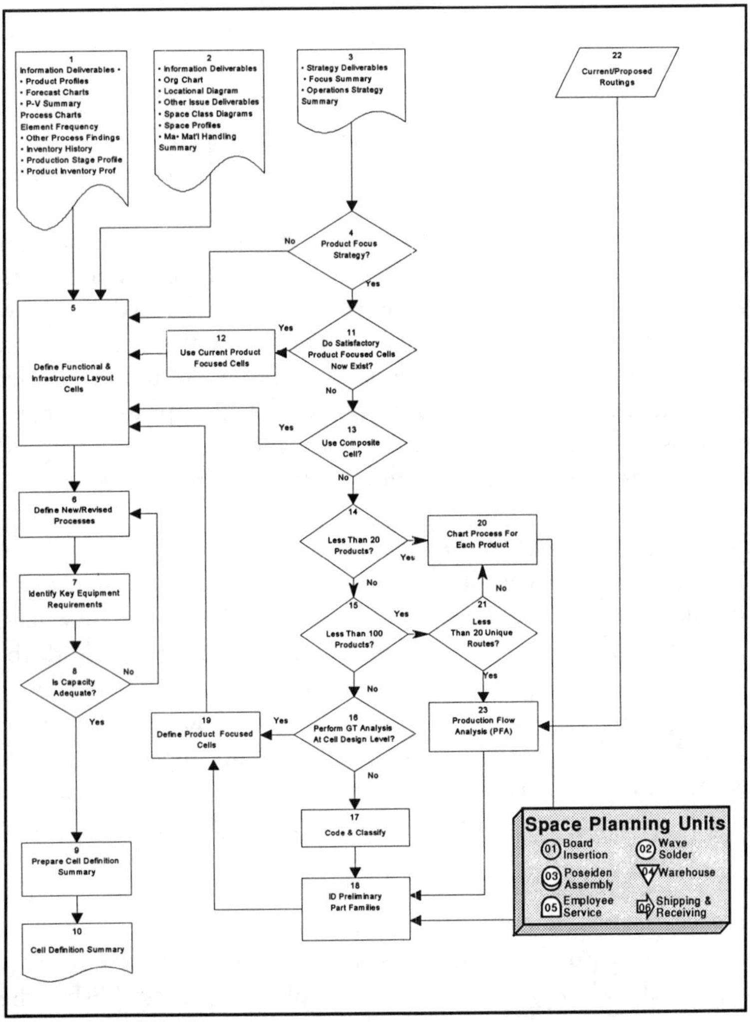

Figure 3.21 - Task 3.11, Define SPUs & Processes

organization chart. We look for activities, people, or equipment that will require space. For each such item, a cell could be defined, or the item could be combined with others into an SPU. Generally, ten to thirty SPUs should be identified.

In the SPU definition summary, the space planners should identify each SPU with a name and number and show those that are included. The space planner may also specify exclusions. The columns

Notes:
1) A/C & SEMI CLEAN ENVIRONMENT
2) GT WORKCELLS IN THIS AREA DEFINED AT LEVEL 05

Approvals:
A. Jackson 11/11/94
C. Hacker 11/12/94
G. Washington 11/12

ID	Cell Name	Inclusions	Exclusions	Source: Existing	Process	Org Chart	GT	Infra	Strategy	R&D	Codes			Notes
01	SILKSCREEN OPERATIONS	SILKSCREEN DRYING BAKING & HUMIDIFY	SCREEN PREP POST-SCREEN OPS	●	●									1
02	SCREEN PREP	PREP STORAGE	SILKSCREEN PRODUCTION	●	●			●						
03	ROLL OPERATIONS (COMPOSITE CELL)	GT CELLS FOR ROLL SLIT & PACKAGE		●	●		●		●					2
04	POST SCREEN OPS (COMPOSITE CELL)	GT CELLS FOR POST-SCREEN OPS	SILKSCREEN OPS		●		●		●					2
05	SLIT & SHEET	ROLL SLITTING SHEET CUTTING	INTERMEDIATE STGE	●	●									
06	INTERMEDIATE STORAGE	WIP FOR COMMERCIAL & ROLL OPS		●	●									
07	RAW MATL STORAGE	ROLL & COMMERCIAL RAW MATLS		●	●									
08	FINISHED GOODS STORAGE	ROLL FINISHED STORAGE	COMML FINISHED STORAGE		●									

Figure 3.22 - Cell Definition Summary

that show the source of the cell definition should be examined (e.g., does this activity show up on the organization chart, process chart, or both?).

If the operations strategy summary calls for a product-focused space plan (line, cellular, or Toyota), support cells still need to be identified. However, a product-focused space plan may absorb many indirect activities within the product-focused cells. For example, in an electronics plant, assembly-integrate-test cells each had one of seven major products. Schedulers, test engineers, and process engineers sat in the cells. A subsequent organization realignment had these people report to cell managers rather than functional managers.

If the plan's strategy calls for product focus, planners should proceed to Block 11 to decide whether a current cell definition is satisfactory. If the factory has previously operated with product-focused cells, rearranging them may be all that is needed. Suitable definitions may have been developed during Task 03.10, "Identify Focus Opportunities." Now, additional product-focused cells should be defined. If the current definition is unsatisfactory, the next step is Block 13.

For a product-focused space plan, the planner can define product-focused cells at the macro- or micro-level. Defining work cells at the macro-level is satisfactory if the likely result is a manageable number and if the effort required is reasonable. Sometimes this is not the case. For example, an adequate definition might require an extensive group technology analysis, which is inappropriate at the macro-level. Or there might be many small cells that are difficult to arrange. If so, the planner should consider using one or more composite cells.

A composite cell consists of several smaller cells. In the Cosmos Products example, post-screen operations and roll operations lend themselves to cellular manufacture. Designing the individual cells and deciding which products go in them is a prolonged, detailed, and difficult task that, in this case, has been postponed until the next design level. Therefore, post-screen operations and roll operations have been defined as composite cells. This was not absolutely necessary. A group technology analysis might have been conducted at this macro-level to identify families and define the subcells.

If composite cells are not used at the macro-level, Blocks 14 through 16 are the next step. Planners evaluate the number of products and select an appropriate analysis tool. For a small number of products, twenty or less, the planners should go to Block 20, chart the process for each, and then use the process charts in Blocks 18 and 19. In Block 18, the space planner identifies preliminary part families; in Block 19, the cells are defined.

For a moderate number of products, less than 100 but more than twenty, production flow analysis is used. The process then moves to Block 18 for defining product families and Block 19 for defining the corresponding cells.

Many products (more than 100) will probably require a classification and coding analysis. This is an extensive undertaking but one with significant benefits.

After defining SPUs, it is time to review the processes for improvement. The process analysis at this level may be general. Examining the process further may be done during the detailing of the layout at the micro-level.

In Block 7, the space planner determines key equipment requirements. This is not always a complete list; rather, it identifies major equipment that occupies significant space or needs significant funding. In Block 8, capacity is checked. Normally, this capacity analysis is confined to key equipment or known bottlenecks. Process charts for any significant process revisions and a list of key equipment might also be helpful.

When complete, a cell definition summary is in place. Cell definition should include every space or feature necessary for the new plant. It is not always an elaborate document. Everyone involved should know what each SPU contains and what it will not contain. These are the building blocks for the new layout.

Because cell definition is so crucial to the remaining activities, planners should circulate it widely for comment and input. In addition, decision-makers must approve it before space planning can go forward.

Cosmos cell and process definition

Part of the cell definition for Cosmos Products is illustrated in figure 3.22. The operational cells come directly from the focus study of Task 03.11. Other cells are derived from the existing process charts and the physical infrastructure checklist.

For example, SPU 01 is silk-screen operations. It includes silk-screen printing, drying, baking, and humidifying and excludes screen preparation and subsequent operations. The team created this SPU from the existing space diagram and also from the existing process chart.

Cell 04 is post-screen operations. It includes thermal die-cutting, pick-and-peel, masking, labeling, inspection, and packaging. The existing space plan had no area with this label. This SPU was derived from the process charts and the strategy statement.

Figure 3.23 is the revised process sheet for Cosmos's multi-color commercial family of products. Comparing this illustration to figure

3.8 reveals that the value added index (VAI) has increased from 0.20 to 0.30. The number of elements has decreased from 78 to 50. Improvements of this magnitude (30 to 50 percent) are not uncommon in layout-reengineering projects. The dotted envelopes in figure 3.23 represent the cells where the process activities occur. Most of the

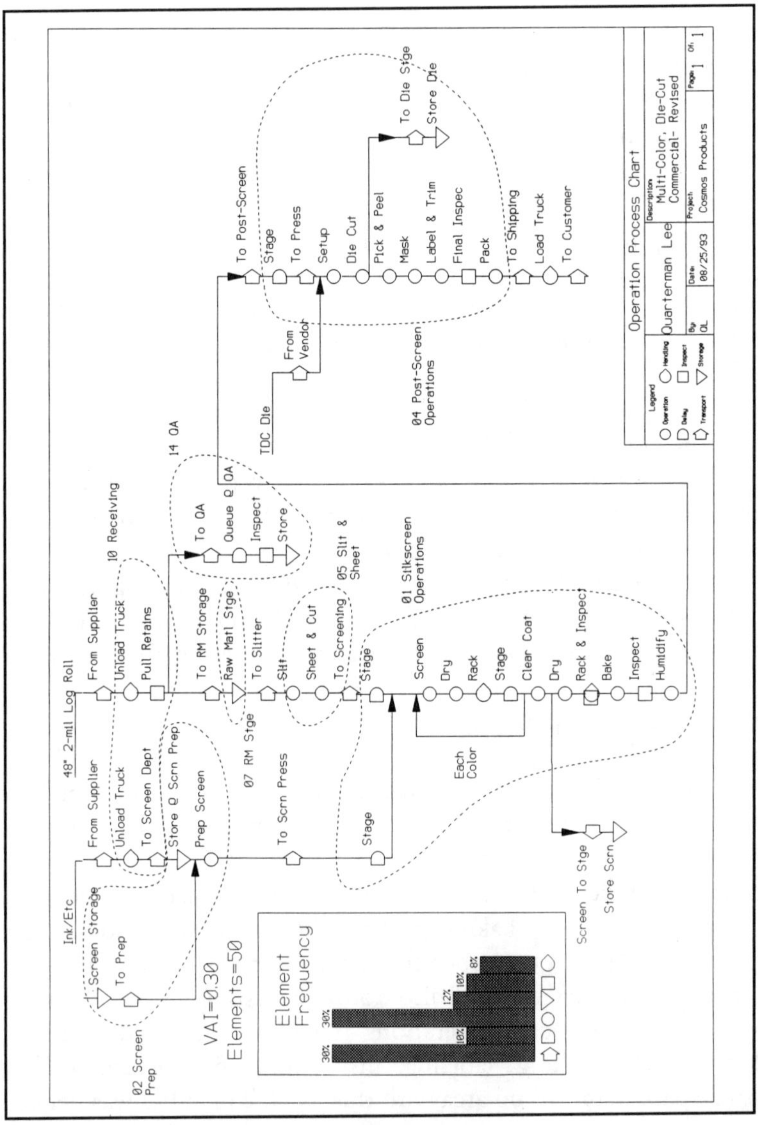

Figure 3.23 - Revised Process For Multi-Color Die Cut Product

process improvements at Cosmos are from the elimination of transport, storage, and delay elements, which are rendered unnecessary when processes occur in the same location. This is the power of a product-focused space plan.

Material flow analysis

In Task 03.12, the analyst uses information gathered earlier to calculate material flow between each combination of SPU pairs. Additional data may be needed for this calculation to establish the affinities associated with material flow. Figure 3.24 shows the procedure for this analysis.

In manufacturing, material flow is usually an important factor in layout. For non-manufacturing space plans, material flow may not be relevant, and this task probably will not apply. Between the extremes, the relative importance of material flow for establishing affinities will vary considerably.

Material flow values are one of two inputs for affinity development. As space plan design progresses and several options are under consideration, the material flow analysis can assist in evaluating these options. Later, flow calculations provide a basis for handling system design.

From the P-V summary, process charts and observation (Block 1) materials are classified into manageable groups (Block 2). This classification assists in developing a common unit for measuring flow, the equivalent flow unit (EFU). A classification summary is one of the deliverables for Task 03.12.

Usually these groups number less than twenty-five. They are based on material-handling characteristics. Structural shapes might be one group in a metal-working factory. It would include steel and aluminum shapes that are ten to twenty feet in length. Another group might be small parts—items defined as less than four ounces and less than three inches on any dimension.

In Block 3 of the procedure, planners choose an EFU. This is a two-part measure: material-units per time-unit such as pallets per day (metalworking); cartons per hour (grocery distribution); tons per day (steel foundry); or totes per day (electronics).

When there is one type of material, this step is easy. It uses the normal unit such as tons or pieces. Such situations are rare, however. Most layouts deal with a wide range of material movement.

Materials also may change form. A sheet metal cabinet for computers begins as a flat sheet that is difficult to handle. Cutting and forming increases the difficulty and bulk by an order of magnitude. Paint makes it delicate and susceptible to damage. Packaging then allows it to be nested and stacked, rendering it less delicate.

Figure 3.25 illustrates these changes in a quantified schematic flow diagram. The lines represent movement, and their width represents the flow rate in equivalent pallets per week. The flow rate in units per week is constant throughout the process. However, the change in size, features, delicacy, and packaging changes the equivalent flow as it moves from one operation to the next.

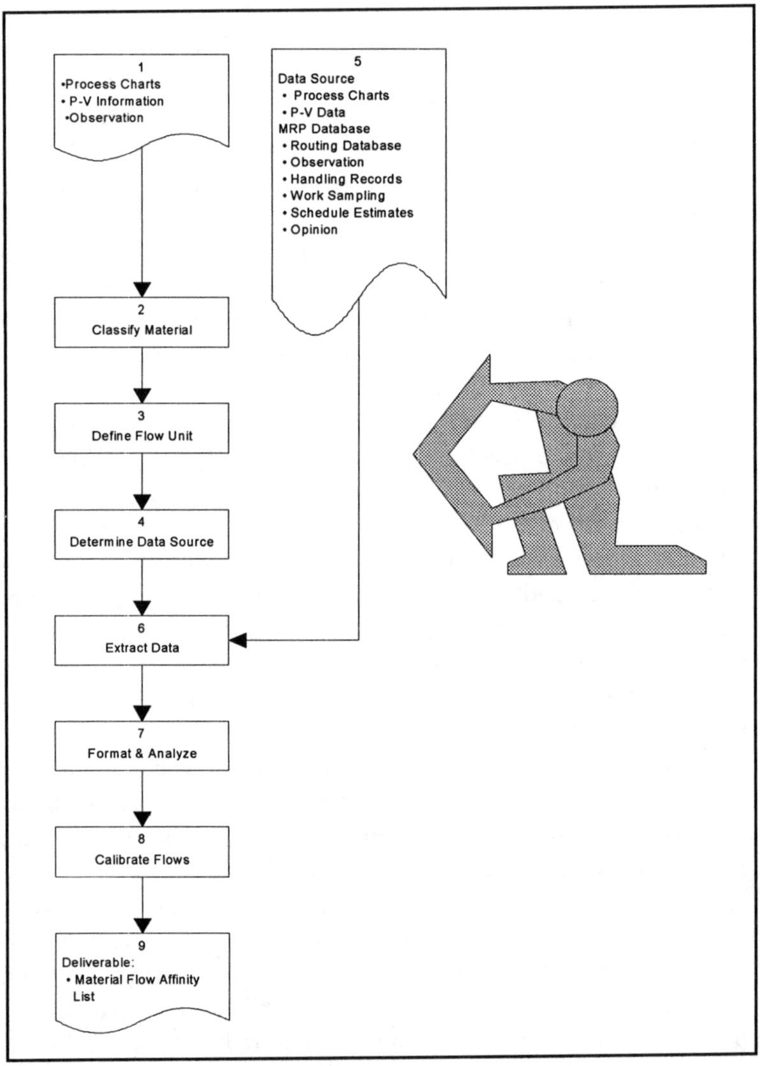

Figure 3.24 - Task 3.12, Analyze Material Flow

The source of data is determined in Block 4 of Fig. 3.24. For simple flow situations, the P-V analysis and process charts provide all the necessary information. In complex situations, the process charts may be too many or too complex; sometimes moves take place that are not in the official process. These must be identified from other data sources such as the MRP database, material handling records, direct

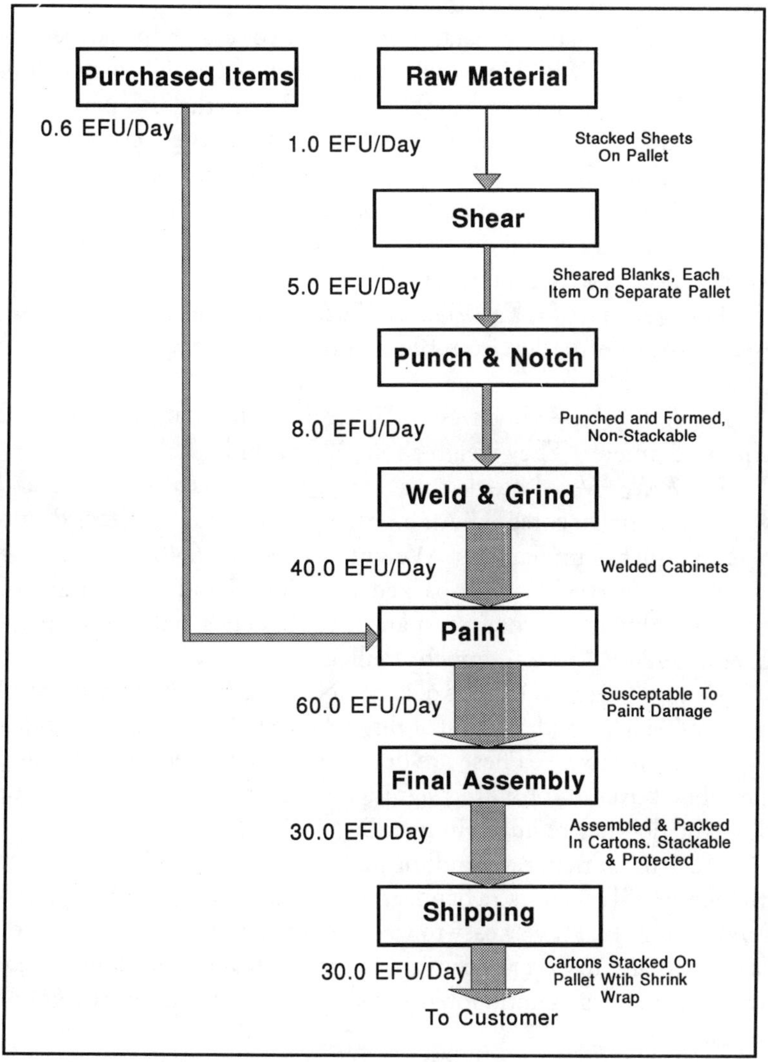

Figure 3.25 - Changes In Material Flow

observation, or random sampling as outlined in Block 5.

Block 6 marks the extraction of data. Again, for simple situations, only the process charts and the P-V information are necessary. For each move on the process chart, the planner determines if a similar move will exist when the new SPUs are used. He or she then determines the flow unit and the number of flow units per day required to meet the sales forecast.

Other data sources may need significant manipulation. Each space plan project is unique with respect to flow data. Experience and common sense are the main guides.

Block 7 formats the data, usually on a computer spreadsheet or database. In Block 8, the flow is calibrated using the AEIOUX conventions (see fig. 2.9). This is done on a ranked bar graph with the SPU pairs along one axis and flow rates on the other. The rating should be done manually. The affinity distribution in figure 2.10 should be used only as a guide, because other factors also are involved. For example, discontinuities in the curve naturally divide one rating from another. Affinity pairs that have zero flow between them get a "U" rating.

The procedure for Task 03.12 is illustrated by the Cosmos project. Based on the information from Block 1, the materials were classified as follows:

Log Rolls-These vinyl stock rolls are 36 inches long and about 12 inches in diameter. They weigh about 200 pounds.

Slit Rolls—Vinyl stock rolls are rolls that have been slit and rewound on smaller cores. They range from 6 to 20 inches in length and less than 5 inches in diameter. Weights are less than 40 pounds.

Roll Packages—These are packed roll products similar to cellophane tape or masking tape. The largest are about 5 inches in diameter and 4 inches in length. Most are much smaller.

Sheets—These large sheets of vinyl stock or decal material average 24 inches by 60 inches and remain flat throughout the process and shipping.

Packaged Sheets—These are decal sheets packaged in corrugated boxes. The boxes and packing significantly increase their volume but reduce the delicacy required in handling.

Much of the material handling in the plant is done using handcarts with four-wheel castors, so a handcart was used as an EFU. This offered several advantages. It was easy to visualize the handcart being used for all materials and to develop conversion factors from that vision. In the new layout, handcarts undoubtedly would remain the primary means of movement.

Table 3.1 is the output of Block 7 and the deliverable for Block 9 on the procedure diagram. This table shows the material flow analysis for Cosmos Products. The material classes are at the top left. Next to

EFU=Equivalent Hand Carts
Log Rolls 0.500
Slit Rolls 0.071
Roll Packages 0.004
Sheets 0.167
Packaged Sheets 0.250

(A) From -To	(B) Units	(C) Fwd Units /Day	(D) Rev Units /Day	(E) EFU Fact	(F) Tot EFUs /Day	(G) Flo Vow Rtg	(H) Flo Num Rtg	(I) Flo/ N-F Ratio	(J) N-F Vow Rtg	(K) N-F Num Rtg	(L) Tot Num Scr	(M) Tot Vow Rtg
01-02						U	0	1.25	A	4	2.00	E
01-04	SHTS	69.0		0.17	11.5	I	2	1.25	A	4	3.25	A
01-05	SHTS		69.0	0.17	11.5	I	2	1.25	E	3	2.75	E
01-08						U	0	1.25	O	1	0.50	O
01-12						U	0	1.25	I	2	1.00	I
01-13						U	0	1.25	I	2	1.00	I
01-14						U	0	1.25	I	2	1.00	
01-15						U	0	1.25	I	2	1.00	I
02-11						U	0	1.25	E	3	1.50	I
03-04						U	0	1.25	A	4	2.00	E
03-05	SLITS		46.0	0.11	5.1	O	1	1.25	E	3	2.13	E
03-06						U	0	1.25	O	1	0.50	O
03-08	PKGS	2134.0		0.01	19.2	A	4	1.25	I	2	3.50	A
03-10	PMAT		7.3	0.25	1.8	O	1	1.25	U	0	0.63	I
03-12						U	0	1.25	O	1	0.50	O
03-13						U	O	1.25	I	2	1.00	I
03-14						U	O	1.25	O	1	0.50	O
03-15						U	0	1.25	O	1	0.50	O
04-09	SHTS	69.0		0.25	17.3	E	3	1.25	E	3	3.38	A
04-11						U	0	1.25	I	2	1.00	I
04-12						U	0	1.25	O	1	0.50	O
04-13						U	0	1.25	O	1	0.50	O
04-14						U	0	1.25	O	1	0.50	O
04-15						U	0	1.25	O	1	0.50	O
05-06	SLITS	46.0	46.0	0.11	10.1	I	2	1.25	I	2	2.25	E
05-07	LOGS		23.4	0.50	11.7	I	2	1.25	I	2	2.25	E
05-09	LOGS	0.2		0.50	0.1	O	1	1.25	U	0	0.63	I
05-13						U	0	1.25	O	1	0.50	O
05-14						U	0	1.25	O	1	0.50	O
05-15						U	0	1.25	O	1	0.50	O
07-10	LOGS		23.4	0.50	11.7	I	2	1.25	O	1	1.75	E
07-12						U	0	1.25	O	1	0.50	O
08-09	PKGS	2134.0		0.01	19.2	A	4	1.25	I	2	3.50	A
08-12						U	0	1.25	O	1	0.50	O
08-13						U	0	1.25	O	1	0.50	O
09-10						U	0	1.25	O	1	0.50	O
09-12						U	0	1.25	O	1	0.50	O
10-14						U	0	1.25	E	3	1.50	I
11-12						U	0	1.25	I	2	1.00	I
11-13						U	0	1.25	O	1	0.50	O
12-13						U	0	1.25	O	1	0.50	O

TABLE 3.1

each class descriptor is the EFU conversion factor, which converts the material unit into an EFU. Each conversion factor is the inverse of the number of flow units that fit onto the cart. A cart usually carries two log rolls, for example, so its conversion factor is 0.5. The SPU pairs are in column A of table 3.1. Column B shows the flow units—slit rolls, sheets, etc. Columns C and D indicate the flow rate.

Paths should be specified using two numerical SPU identifiers. In the Cosmos example, 03 refers to roll operations, and 06 refers to intermediate storage. The flow path between them is 03 to 06. To avoid duplication and possible errors, planners should use only the forward path—the SPU with the lowest number followed by each SPU numbered above it. When material moves from a higher numbered SPU to a lower numbered SPU, it is called reverse flow. The total material flow is the sum of the forward and reverse flows. Column F in the Cosmos model is the flow totals multiplied by the EFU factor. This result is the average flow rate in EFUs per day. Column G shows the vowel rating for each flow path.

Figure 3.26 shows the flow calibration for Cosmos Products. This is Block 8 on the procedure chart and is typical of a product focused layout. It has a small number of high flow rates and many SPU pairs with zero flow. Process-focused layouts have a much broader distribution of flows commensurate with their complex natures.

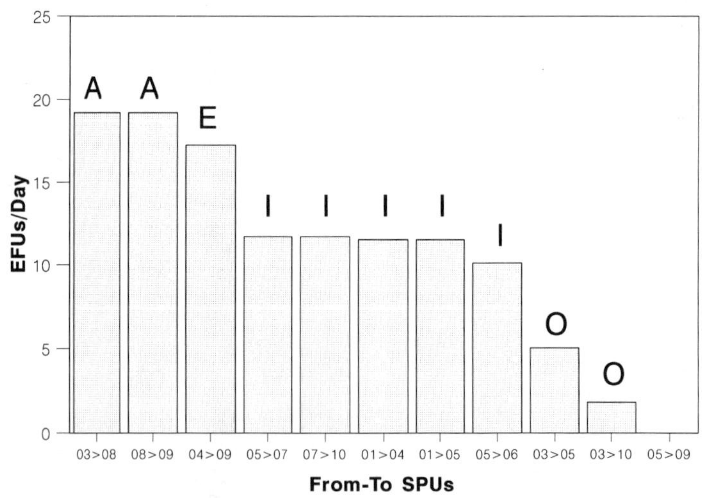

Figure 3.26 - Material Flow Calibration

Identifying non-flow affinities

Material flow is only one of many factors that give rise to affinities. Other factors are intangible and more difficult to quantify. Examples of these factors are: personal communication; the need to transfer personnel between cells or departments; movement to and from the cafeteria or rest rooms; quality feedback; joint teamwork communications; access by outside visitors; RF communications requirement; and other site-specific needs.

Figure 3.27 shows a chart for recording non-flow affinities that also may be used to document flow affinities or total affinities. Diagonals represent each SPU. When they cross, they form a diamond. In the upper half of the diamond, the affinity rating is recorded using the vowel or

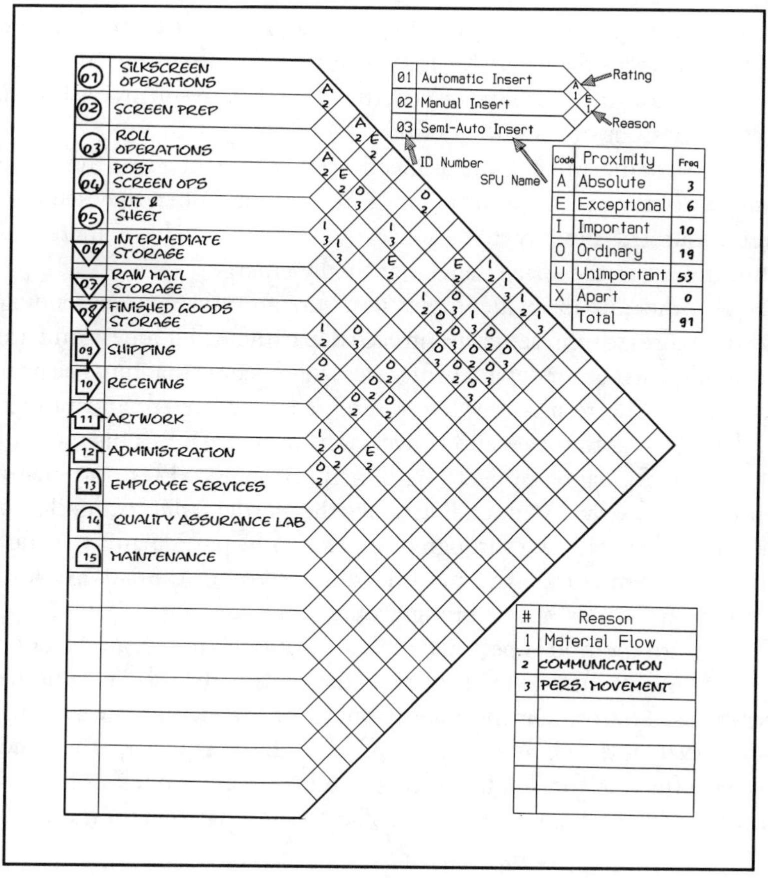

Figure 3.27 - Affinity Chart

number scale shown in figure 2.9. The lower half of the diamond is the place to record the primary factor(s) that gave rise to the affinity.

These non-flow affinities are independent of material flow requirements. The problem lies in capturing them. In Task 03.14 they are merged with affinities for an overall or total affinity rating. A survey, consensus meeting, or personal evaluation may also be used.

A consensus meeting that assembles representatives from each department or SPU is usually the best approach for accomplishing Task 03.13. The analyst acts as facilitator. Using the affinity chart of figure 3.27, he or she explains the need for affinity ratings, the chart, and the desired distribution. The group considers each pair of SPUs, one at a time, and discusses the relationships.

Using the conventions in figure 2.9, they decide on a rating. Initially, these discussions are rather long. After five to ten ratings, however, the group will begin to agree readily. A scribe records the ratings and keeps the group focused by displaying the current SPU pair on cards. Frequently, corollary issues arise. These may result in constraints or even revisions of the SPU definition.

Participants in a consensus meeting emerge feeling that they have been part of the overall project. This is important. When they see how their input led directly to a space plan, they will have increased commitment to the space plans that finally emerge.

Another method of identifying non-flow affinities involves sending questionnaires to representatives in each department. The questionnaire asks them to list other departments, areas, and people that must be near each other. The results are then assembled, interpreted, and ratings developed using the scale and conventions in figure 2.9. This method is effective for large projects with fifty or more SPUs and many affinities. However, it does not allow the participants to develop a common understanding through discussion. The participants may not trust the judgment of the person who interprets the surveys and corollary issues may not be brought out.

A third method is personal evaluation, which uses a single judge to determine affinities. He or she must have intimate knowledge of the operations. This often is the analyst, but he or she may also be a strong leader, perhaps the plant manager or CEO. This is a quick method and may be effective for small projects. This, however, does not build consensus and may be divisive. Corollary issues may remain hidden.

Merging affinities

Two sets of affinities now exist. The development of flow affinities used a quantitative approach. Non-flow affinities by their nature preclude a quantitative approach and were identified by a consensus or some other non-quantitative approach. These must now be merged into a single set of affinities (fig. 3.28). This is Task 03.14.

A spreadsheet created by hand or computer is usually the most straightforward method of merging. Table 3.2 is an extension of the spreadsheet in table 3.1. These columns are put in after columns A through D.

Column E: Vowel Non-Flow Rating (Enter Manually)
Column F: Numeric Non-Flow Rating (Enter Manually)
Column G: Flow/Non-Flow Ratio(Enter Manually)
Column H: Merged Score: Col. A x Col. F + Col. E x (1 - Col. F)
Column I: Merged Vowel Rating (Enter Manually)

Planners should add rows for all remaining combinations of SPUs, sort the rows in the merged score column (Col. H) in descending order, and plot the merged scores on a ranked bar chart. From the chart,

EFU=Equivalent Hand Carts

Log Rolls	0.500
Slit Rolls	0.071
Roll Packages	0.004
Sheets	0.167
Packaged Sheets	0.250

(A) From —To	(B) Units	(C) Fwd Units/ Day	(D) Rev Units/ Day	(E) EFU Fact	(F) Tot EFUs/ Day	(G) Flo Vow Rtg	(H) Flo Num Rtg
03-08	PKGS	2134.0		0.01	19.2	A	4
08-09	PKGS	21234.0		0.01	19.2	A	4
04-09	SHTS	69.0		0.25	17.3	E	3
05-07	LOGS		23.4	0.50	11.7	I	2
07-10	LOGS		23.4	0.50	11.7	I	2
01-04	SHTS	69.0		0.17	11.5	I	2
01-05	SHTS		69.0	0.17	11.5	I	2
05-06	SLITS	46.0	46.0	0.11	10.1	I	2
03-05	SLITS		46.0	0.11	5.1	O	1
03-10	PMAT		7.3	0.25	1.8	O	1
05-09	LOGS	0.2		0.50	0.1	O	1

TABLE 3.2

they should assign a merged vowel rating (Col. I).

Deciding the relative importance of flow and non-flow factors depends on the industry, process, and other influences. Heavy industries such as steel or shipbuilding warrant a flow/non-flow ratio up to 2.0. Office areas and industries that depend heavily on personal contact may have ratios as low as 0.5.

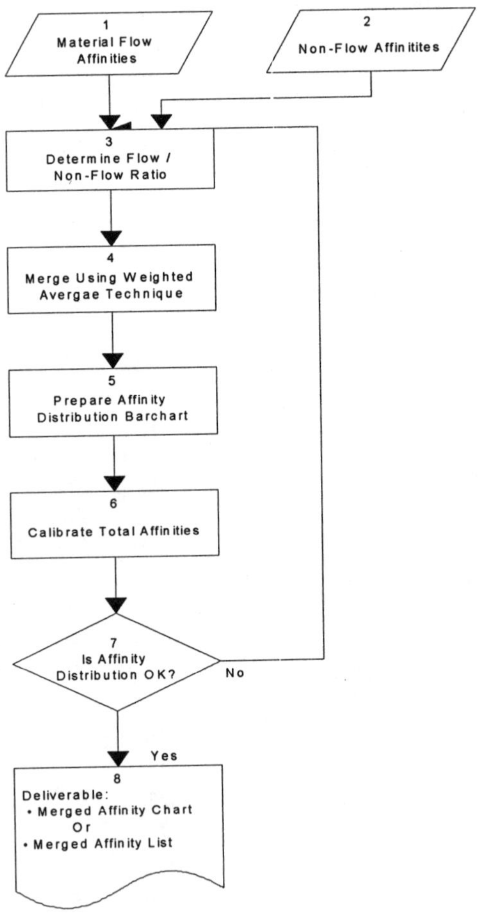

Figure 3.28 - Task 03.14, Merge Affinities

Generally, the same flow/non-flow ratio should be used for all the affinities on the space plan. Occasionally, however, specific affinities may have to be modified for special circumstances.

Next, the total affinities must be rated. When assigning the merged vowel rating, the analyst should consider two factors. First, he or she should strive for a workable distribution of ratings similar to those shown in figure 2.9. In addition, a search for natural breaks or discontinuities in the distribution avoids having nearly identical scores with different ratings. Accuracy is not paramount in this process.

Developing a configuration diagram

In Task 03.15, merged ratings are used to develop a configuration diagram. The configuration diagram is the first of the derived elements. It comes from cell definitions, affinities, and experience. The graphics work may be done on a CAD system or other software. However, manual development is straightforward and often quicker.

To develop the affinity diagram, the analyst places the A affinities and their associated SPU symbols first, then adds the E affinities. At this point, rearranging the diagram is desirable. Next, the I affinities are added and the diagram is rearranged again. He or she finishes with the O affinities, which usually will have little effect on the diagram..

Striving for short distances between the As and Es with minimal crossing is a worthwhile goal. Multiple crossings might create traffic congestion on the final space plan. Lower value affinities probably will have longer distances. The high value A and E affinities will have short distances. Attempting to fit this diagram into a building environment at this point is not advised. some excellent arrangements may be overlooked.

Figure 3.29 illustrates the diagram developed for the Cosmos project. Step 1 features the SPU symbols. Step 2 shows the A and E affinities in an undesirable arrangement. Step 3 shows the rearrangement of the A and E affinities and the addition of the I affinities. Another rearrangement is illustrated in Step 4. Finally, in Step 5, the affinities have been rearranged again, and the O affinities added.

The Cosmos configuration diagram in figure 3.29 is only one of many possible diagrams that uses this combination of SPUs and affinities. It can be mirrored or rotated. There may be other positions for the SPUs that give the same or better results. Some of these variations will fit the building better than others. However, it is best not to jump ahead and anticipate the shape of the building. It is worthwhile to ask several people to develop diagrams, thereby ensuring a wide selection of possibilities.

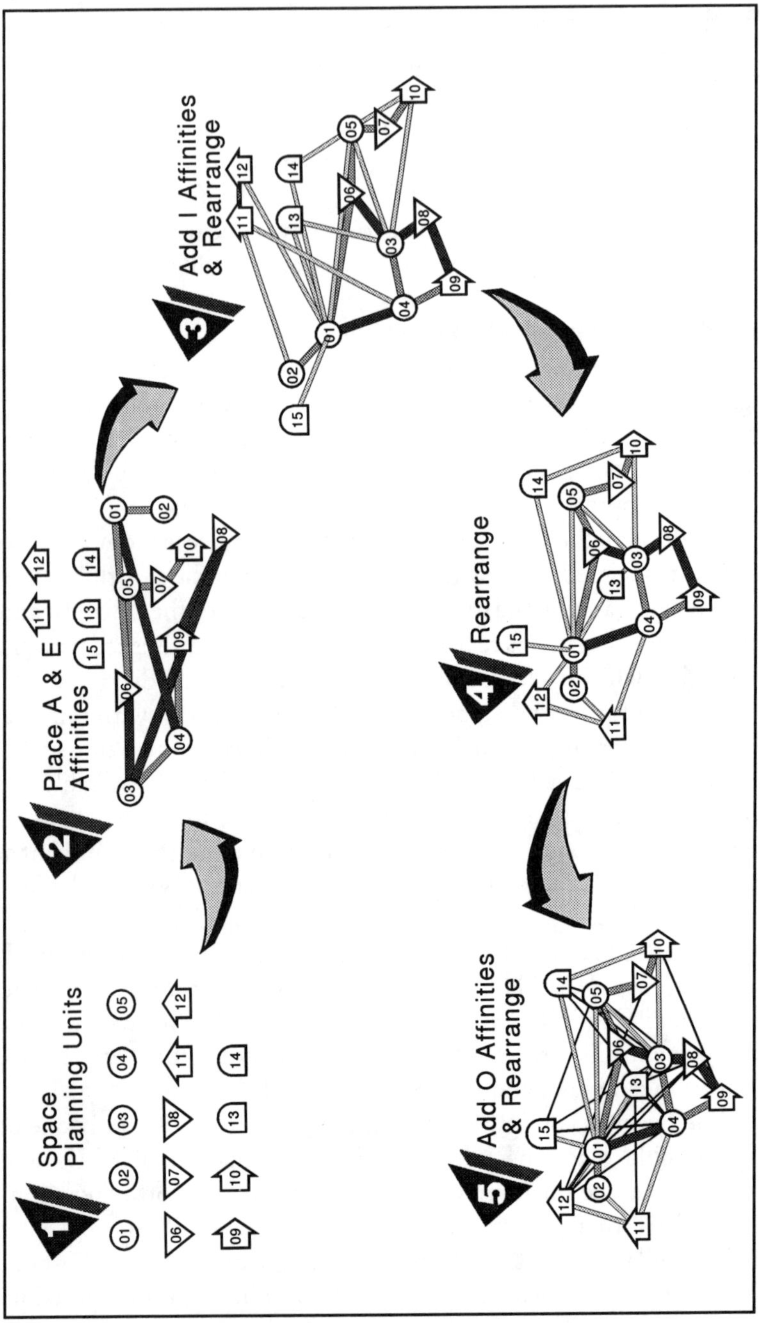

Figure 3.29 - Affinity Diagram

Space calculation

Space is the third fundamental element of a space plan. It is a limited resource; there is only so much space under a roof, on a site, or in a department. Whether the space is land on a site or space in a building, it is usually expensive

Although space is three-dimensional, most space plans ignore the vertical dimension. This is acceptable in all but a few situations.

Most layouts attempt to optimize the use of space as well as its arrangement. A complete space plan requires not only the location of SPUs, but their size and shape as well. The space occupied by SPUs usually prevents the designer from honoring all affinities

ID	Space Planning Unit	1994 Space	1998 Space	Calculation Method
01	Silkscreen Ops	16425	15580	Transform
02	Screen Preparation	1487	1500	Transform
03	Roll Operations	6123	8650	Transform
04	Post-Screen Ops	5460	7855	Elemental
05	Slit & Sheet	909	1000	Elemental
06	Intermediate Storage	3611	1600	Transform
07	Raw Mtl Storage	5151	4800	Transform
08	Finished Goods Stge	6789	4900	Transform
09	Shipping	874	1050	Transform
10	Receiving	1299	1500	Transform
11	Artwork	6320	6455	Standard
12	Administration	7320	8710	Standard
13	Employee Services	714	1500	Transform
14	Quality Assurance	480	745	Elemental
15	Maintenance	558	600	Transform
16	Aisles	2987	3700	Proportion
		62782	65845	

TABLE 3.3

simultaneously. Between them, it forces compromises above and beyond those arrived at in the configuration diagram. The space needs of SPUs may distort even a nearly perfect diagram.

Task 03.16 calculates size of the required space for each SPU, usually in square feet or square meters. Calculation of space requirements uses one or more of six methods. These methods are: *elemental calculation*, *visual estimating*, *transformation*, *space standards*, *proportioning*, or *ratio forecasting*. Table 3.3 shows Cosmos Products' space requirements and how the analysts used several methods for the calculation.

Elemental calculation

This method, illustrated in figure 3.30, starts at the most detailed level. Each piece of furniture or equipment assigned to an SPU is measured. These dimensions are then added together for the total amount of space. Space for aisles, miscellaneous storage, or other needs are also included in the sum. This added space often is a percentage of the basic equipment space.

Elemental calculation is simple and straightforward. However, it has its limitations. For one, it takes considerable time and effort. Uncertain forecasts can make it difficult to determine how much furniture or equipment will occupy the space. Elemental estimating is primarily a short-term methodology. Most industries use it for one to three years in the future. Beyond that, other methods are equally and perhaps more reliable.

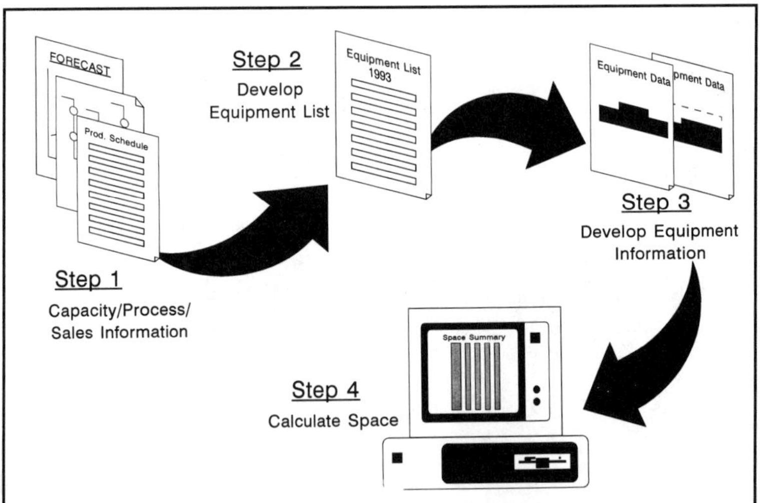

Figure 3.30 - Computing Space, Elemental Calculation

Visual estimating

Visual estimating uses templates to represent equipment and furniture—often this is only primary or large-scale equipment. The designer places the templates on a layout according to his or her judgment and knowledge of similar installations. The arrangement need not resemble the final configuration. The designer adjusts for aisles; maintenance access; and materials, storage, and minor equipment space requirements, and then measures the total space required for that SPU. Figure 3.31 illustrates.

Visual estimating is very useful for an SPU dominated by a few large items, but it can be laborious when many small items occupy the space. Forecast accuracy may become difficult. In the hands of an experienced planner, it can be accurate; however, inexperienced planners should be very cautious with this method.

Transformation

Figure 3.32 illustrates the transformation method: a designer takes a space requirement for an existing SPU and uses that as the basis for projecting future needs. Adjustments must be made for existing conditions such as overcrowding or wasted space. Changes in production and product mix are additional considerations, and new technologies that may need more or less space than current technology should be examined.

For example, suppose that current space for work in process storage is 3,400 square feet. Observation indicates that the area is crowded and congested. Ten percent is added to bring the space allocation up to present-day requirements. Then, if the planning

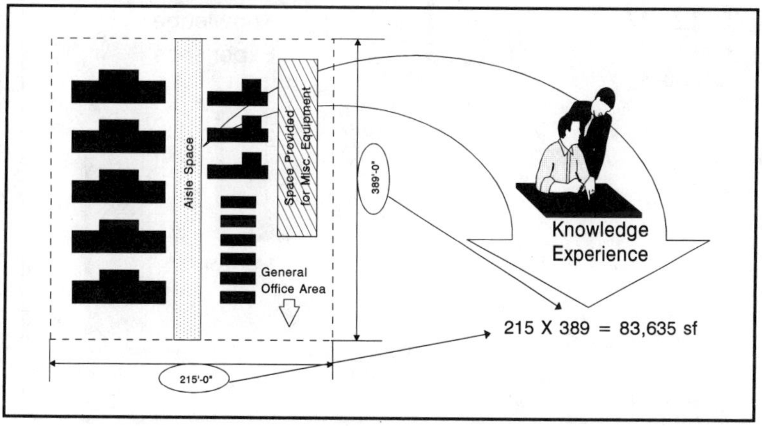

Figure 3.31 - Computing Space, Visual Estimation

horizon has a production increase of 22 percent, this would usually call for a corresponding 22 percent space increase. However, the firm also plans to improve production control and expects a 13 percent overall reduction in space. The calculations are: 3,400 x 1.10 x 1.22 x (1.0-0.13)=3,970 square feet.

Transformation is versatile. For an experienced space planner, it is reliable and accurate. It does not require detailed knowledge of equipment and furnishings.

Space standards

Space standards are useful in repetitive situations. Based on experience, space requirements that relate to some parameter of the business are established. For example, the United States Postal Service allocates space for certain operations based on the number of mail pieces handled per day. Many firms use a standard space and workstation design according to job classifications or places in the hierarchy.

Standards call for history, unchanging situations, and detailed knowledge or forecasts. They are common in office layouts. Each department receives space based on the number and positions of people assigned to it. Figure 3.33 shows the process, and Chapter 7 has examples.

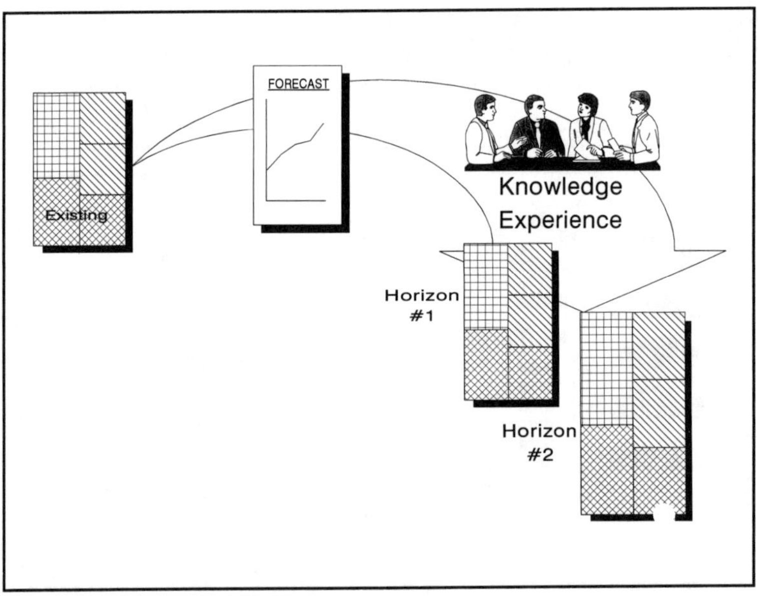

Figure 3.32 - Computing Space, Transformation

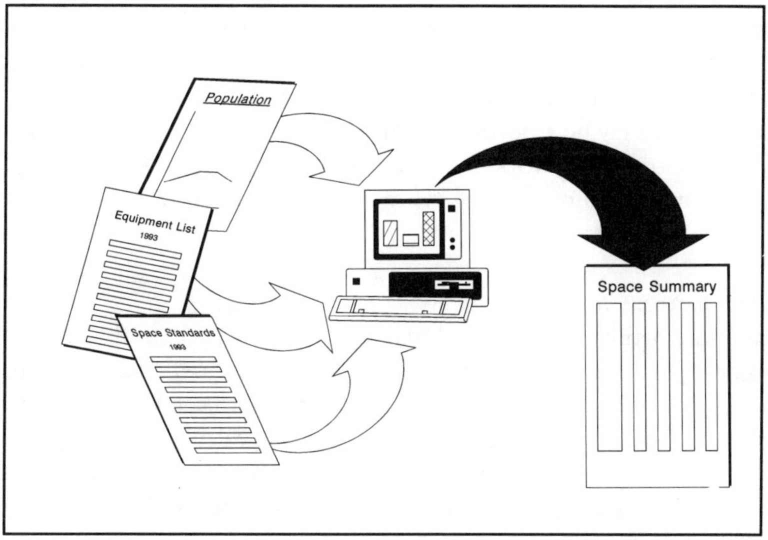

Figure 3.33 - Computing Space, Standard Data

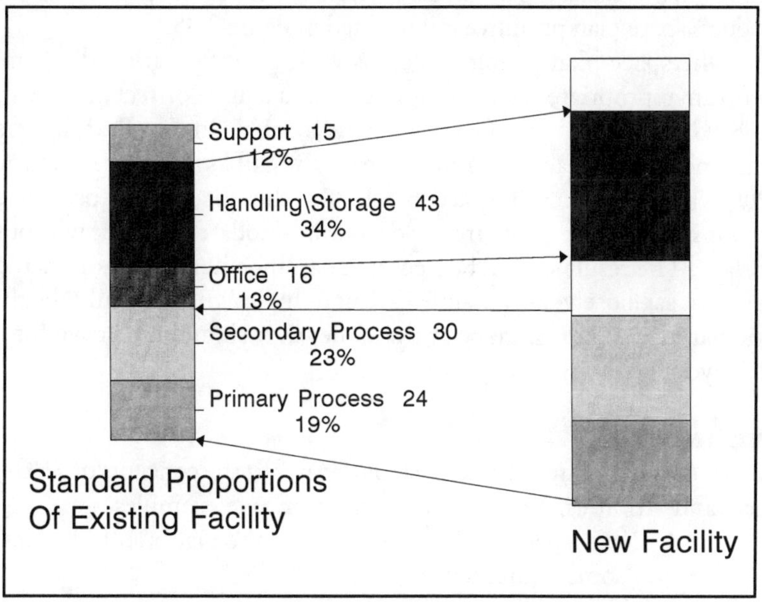

Figure 3.34 - Computing Space, Proportioning

Proportioning

Certain types of space calculation use proportions effectively. The space for a given SPU comes from the calculation of another space. For example, aisles might be a percentage of production space, or conference room space may be a portion of office space. Figure 3.34 illustrates.

Proportioning works well when the history to support it exists. It usually applies to only a few types of space, however. Proportioning requires little effort.

Ratio forecasting

Ratio forecasting uses historical trends to forecast space. In this method, business parameters and space are correlated over time. Such ratios may change gradually over the years. The analyst then projects the trend of this ratio into future years and uses that projection to calculate space.

Ratio forecasting, which is based on historical data, is most appropriate for long-term site plans. It has limited use for short-term space calculations.

The space plan primitive

The next step in the progression is the space plan primitive, which involves adding space to the configuration diagram(s). The space requirements come from the calculations and space summary (Task 03.16). The Cosmos Products space plan primitive is illustrated in figure 3.35.

The space plan primitive begins with a configuration diagram. Using an appropriate scale, designers place a square or rectangle with the SPUs calculated area near each SPU symbol. In step 2, designers either move each space block underneath the SPU symbol or move the SPU symbol over each space block. As the space plan primitive develops, it will have to be stretched to accommodate the space without overlaps. The result should be a compact arrangement that honors the affinities as closely as possible. Although designers should begin anticipating a building shape at this time, they should not strive for a final layout.

Constraints

Many factors that affect a macro-layout do not fit the concepts of SPUs, space, and affinities. These are constraints. Some examples are:

- Column spacing of 32 feet restricts the placement of aisles and some equipment.
- High electrical load restricts the placement of heat treat ovens to certain areas with adequate electrical service.
- A cold climate dictates that dock doors should not have

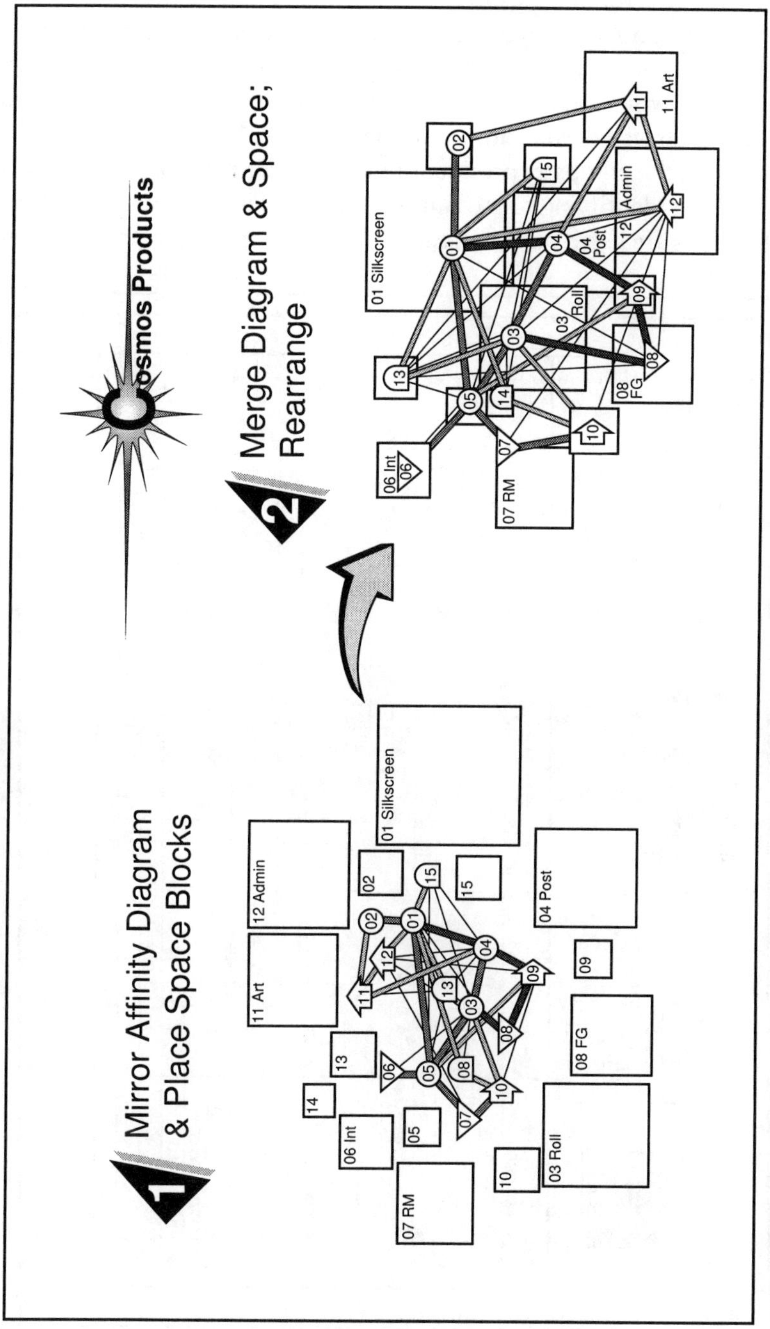

Figure 3.35 - Cosmos Spaceplan Primitive

northern exposure.

- zoning requirements specify that docks not face the street;
- floor loading restricts the placement of certain equipment;
- explosion hazard dictates that a hazardous chemical room have an explosion vent on an outside wall; and
- the company president requests a window for his office.

A form for identifying constraints (Task 03.19) is shown in figure 3.36. The SPUs are listed on the left and across the top, major categories are

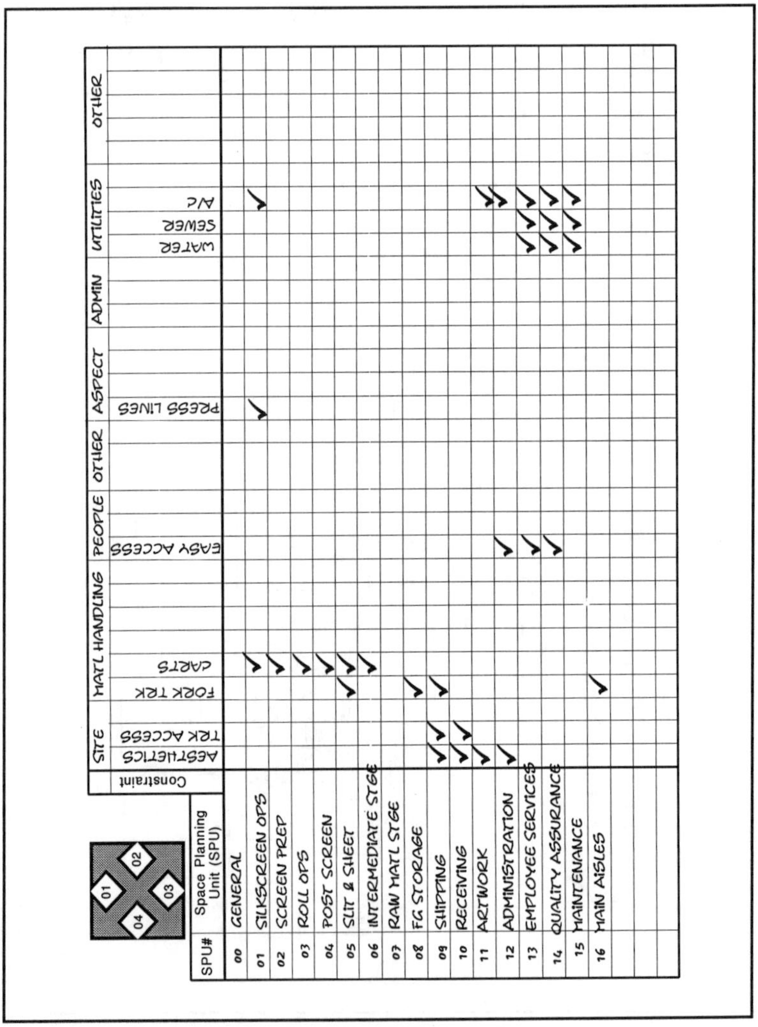

Figure 3.36 - Constraint Summary

identified. These include: site conditions, utilities, handling methods, personnel, procedures and controls, shape ratio, and others. The accumulated project documentation for each SPU and category are reviewed, and the constraints are listed. A bullet or check associates each constraint with a particular SPU. Some constraints apply to all SPUs. In this situation the bullet goes in the "general" row. The following is a description of Cosmos Products constraints:

Aesthetics—The nature of Cosmos products is artistic. Therefore, the aesthetics of the building and surroundings is important. Cosmos wants to present itself well to customers and other visitors. A pleasant surrounding will help attract the best commercial artists. The aesthetics issue applies to shipping and receiving areas, which are often unsightly. It also applies to the artwork, administration, and employee service areas.

Truck Access—This is another site condition. Trucks need access to both shipping and receiving. Cosmos is fortunate in this respect. The site has good access on both the north and south sides.

Fork Trucks—Handling in some areas uses forklift trucks. Adequate aisle widths on the main aisles and selected departmental aisles are necessary.

Carts—Small carts convey materials in many areas. Here, narrower aisles will suffice.

Easy Access—This is a personnel issue. Administration, employee services, and maintenance all require easy and inviting access for people.

Press Lines—Aspect ratio refers to the relative length and width of SPUs. Silk-screen printing operations require a minimum length to accommodate the long press lines.

Utilities—Certain SPUs require water, sewer, and air conditioning.

Next, this constraint summary and the space plan primitive will be used to prepare space plan options.

Designing macro-space-plans

The space plan primitive now must fit into a building outline. The building may exist or it may be a proposed structure.

Preparing space plan options begins with overlaying the building outline with a space plan primitive. The space blocks are shaped to fit building walls, columns, and other features. The constraints summary should be consulted during the placement of each SPU.

For each space plan primitive, there probably will be several viable layouts. All variations of the primitives, including mirror images and rotations, should be examined.

It may be difficult to match space, honor constraints, and design an orderly arrangement. In general, designers should strive for clean, rectangular areas. Space requirements may have to be compromised.

The original space calculations are usually flexible within a reasonable range—10 to 20 percent.

One of the macro-space-plan options (Option 1) from the Cosmos project is illustrated in figure 3.37. This option is based on the operations strategy, which called for a product-focused space plan using cellular manufacturing techniques and focused factories.

The space between the two original buildings is now enclosed to accommodate the increased space requirements and also to allow

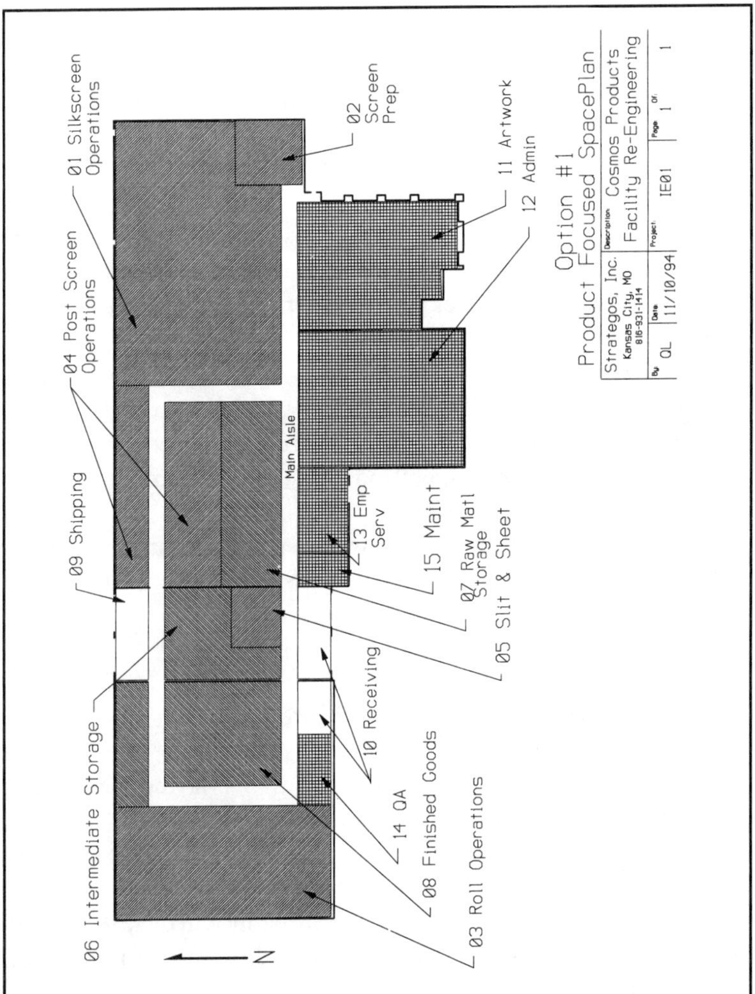

Figure 3.37 - Option 1 of the Macro-Spaceplans for the Cosmos Project

improved material flow between the buildings and to the shipping and receiving docks. Each of the old buildings has become a semi-focused factory, with roll products on the left and commercial products on the right. In the center are slit-and-sheet operations, which serve both focused factories. Service facilities such as quality assurance, employee services, and maintenance are also in a central location. Artwork, administration, and employee services face the street. This satisfies the aesthetic constraints.

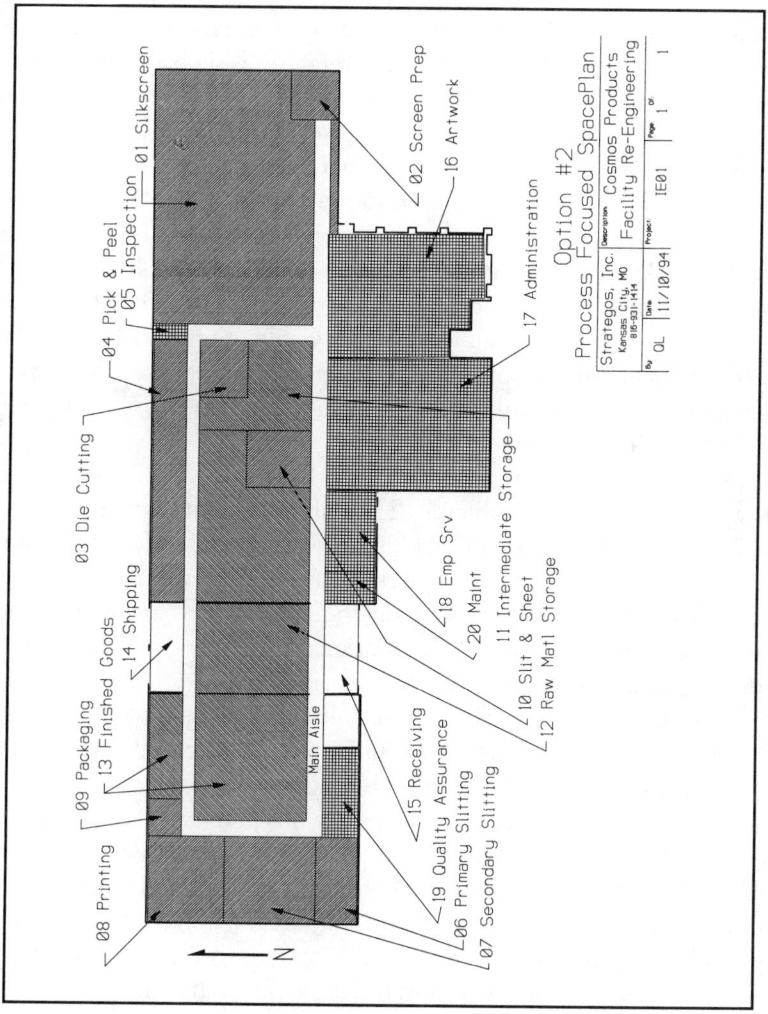

Figure 3.38 - Option 2 of the Macro-Spaceplans for the Cosmos Project

Some Cosmos managers had reservations about the product-focused strategy adopted in Task 03.10. Therefore, an additional macro-space-plan based on continuing the process-focused approach was prepared. The result was Option 2 (fig. 3.38), which mitigated their concerns and demonstrated the relative advantages and disadvantages of process and product focus. It was prepared as a second, parallel project starting from Task 03.10. The process-focused strategy statement produced a different set of SPUs, affinities, and space requirements.

Option 2 retains the aisle system and many of the good features of Option 1. Functional and semi-functional areas such as shipping, receiving, and artwork have many of the same characteristics. They are often in the same location. Some functional SPUs such as storage areas need significantly more space. Process areas change their names, characters, space requirements, and other characteristics. Option 2 needs about 10 percent more space than Option 1. This additional space is in a building expansion on the east side.

For most macro-space-planning projects, there will be three to six fundamentally different options and several variations. Using the existing layout—or simply doing nothing—is always an option. Even when the existing space plan is no longer viable, it makes a convenient baseline for gauging improvement. The Cosmos project team developed several other options, which are not included in this book.

Aisles

Aisles present special problems. They should be straight and wide enough for two-way traffic. Usually, the best approach identifies main aisles as a separate SPU. Designers then place them on the macro-space-plan. Departmental aisles, on the other hand, are within the space calculations for each SPU. Aisles adjacent to walls are often undesirable because they serve only one side.

An alternate approach includes all aisles as part of the SPUs. Designers then place SPUs on the plan, recognizing that those main aisles generally will follow the SPU boundaries.

The straightforward Cosmos space plans in figures 3.37 and 3.38 use the first method. Main aisles have a separate calculation and every SPU is adjacent to a main aisle. A central loop allows continuous traffic in both directions. A single dead-end aisle serves screen prep and part of the silk-screening operation. This aisle system will allow subsequent layout changes without disturbing the basic flow pattern.

Identifying key material handling issues

Material handling and layout are intertwined. The best handling system depends on the space plan and the best space plan may depend on handling methods. Often, a layout that does not work with manual handling becomes viable with automated or conveyorized handling.

This presents a chicken-or-the-egg problem. Are handling equipment and containers selected before the layout? Is the layout designed first and the handling system then selected? Usually, the best approach is to design the layout assuming conventional, perhaps manual handling. This optimizes material flow and often eliminates the need for complex and expensive handling systems.

However, particular handling issues that drastically affect space plan selection must be identified. For example, a pneumatic transport system has different requirements than a system that uses forktrucks for conveying bulk material. One space plan might be the best for fork truck handling, while another might be best for the pneumatic system.

A space plan should be designed and selected before the handling system is finalized. To do this, however, may mean assuming a general type of handling system prior to layout design.

To accomplish Task 03.20, examine each of the proposed layout options and ask the following questions:

- What types of handling systems are viable for each option?
- Would a particular handling system affect one layout option more than another?
- Would a different handling system allow new layout options?

If a particular handling system affects all layout options equally, selection of that system is not a key issue. In such a case, the evaluation of the options is the next step in the space plan. If a particular handling system would give one option a significant advantage over the others, such selection is a key issue. In such a case, further investigation is recommended, perhaps accompanied by a preliminary design and cost estimate for the handling system. This approach allows management to select the best layout at the macro-level without completely designing handling systems for all the options.

Deciding on the best space plan

Several viable options now exist for the macro-space-plan. Many others probably have already been screened out during earlier parts of the design process. The designer should narrow the choice to three to six significantly different options. Each option may have several minor variations. Management and others involved in the project then decide which to use. This is done for several reasons:

- management often criticizes the engineering staff for tunnel vision. Engineers may lock onto an idea early in a project. A wide variety of options shows that the designer or design team has considered a wide range of possibilities;

- asking for a selection from among options is usually more palatable than asking for approval of a preordained design;
- the process of decision-making builds consensus, support, and confidence. This prevents later attempts at redesign by those who felt left out of the process; and
- the decision process may generate hybrid plans, which are often superior to the original designs.

Figure 3.39 shows the procedure for evaluation. The space plan options are the "input" listed in Block 1. Block 2 assembles a decision team.

In Block 4, the team reviews the project's original objectives. These original objectives may be specific, directly measurable, and

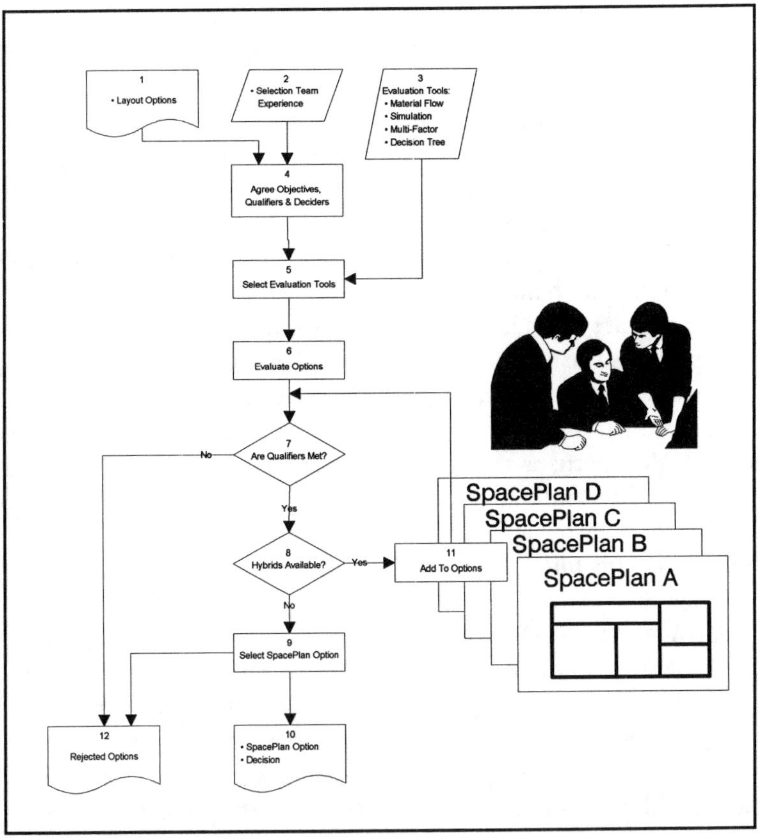

Figure 3.39 - Computing Space, Visual Estimation

applicable to the evaluation, or they may be global and difficult to measure. They may require sub-objectives for a good evaluation. The information developed during information gathering may have modified the objectives. The debates during the strategy development also may have changed objectives. The decision team adopts the original objectives or revises them as appropriate.

Flowing from the objectives are decision criteria. These are factors that the team can evaluate directly-either qualitatively or quantitatively. They are the basis for the decision. Examples of decision factors are: material handling savings, improved communication, OSHA compliance, improved teamwork, initial cost, operating cost, quality enhancement, improved delivery reliability, improved delivery speed, and ability to use a particular technology.

A decision criterion may be a decider or a qualifier. Qualifiers are go/no-go criteria: a space plan meets the minimum requirements or not. Performance beyond the minimum creates no additional benefits. Performance below the minimum disqualifies the space plan from consideration. For example, OSHA compliance might be a qualifier. Layouts that meet the requirements are acceptable. Layouts that go beyond OSHA requirements bring no perceived additional benefit.

Deciders bring additional benefits for each increment of performance. Improved cost, for example, is usually a decider. Option B may have an operating cost advantage over Option A. Although both space plans meet the budgeted cost improvement objectives, Option B is the preferred space plan on that dimension.

Block 3 contains the tools for evaluation. In addition to macro-space-planning, these tools apply to other levels of facility planning. Among the common tools for evaluation are: material flow analysis (MFA), financial analysis, ranking, instinct, positive-negative-interesting (PNI), decision tree analysis, and weighted factor analysis.

MFA examines the large-scale material movement between SPUs. It develops a measure of associated cost and difficulty. Improved communication and coordination are corollary benefits of improved material flow. Specific techniques in this category include transport work, flow diagrams, and D-F plots. For the most part, these are quantitative methods.

Financial analysis includes cost estimating, return on investment (ROI), and payback. These methods are quantitative; however, they often involve qualitative judgments as well.

A simple *ranking*, from most preferred to least preferred, is often an effective tool. The ranking can use qualitative factors, quantitative factors, or both.

The gut-level reaction or *instinct* of knowledgeable people has value. Although it should rarely be used as a primary evaluation tool, it may uncover unseen opportunities or problems.

PNI analysis is a variation of the brainstorming technique. It examines each space plan factor, focusing first on the positive features and then on negative features. Finally, it focuses on those that are neither positive or negative—things that are interesting or unique. This analysis, was developed by Edward DeBono, an expert on thinking processes. It is simple but effective. It often brings out hidden features and builds teamwork and consensus.

Decision tree analysis is useful when a series of probable events can affect the decision. For example, which space plan is best if a particular contract is won and, afterward, the overall market contracts? It helps evaluate the cumulative probability of each of the four possible outcomes. Combined with financial analysis, it is a quantitative tool.

Weighted factor analysis bases a decision on a combination of the various factors, both qualitative and quantitative. It is best if the factors are independent, but this is not always possible. Some compromising of this principle is acceptable. Judges first identify the factors, then decide a weight for each, and, lastly, rate each option.

In addition to the tangible and intangible categories, strategic issues may arise. These are usually qualitative. The consequences of strategic issues are often so far-reaching and so important they overshadow all other factors. For example, Option A might use a new technology. This technology shows no immediate cost benefit yet the introduction potentially could revolutionize the industry and place a firm far ahead of others. Should Option A be selected? This is a decision for top managers and cannot be made lightly.

In weighted factor analysis, the judges that weigh the factors may be different from those who rate the options. For example, top management may weigh the factors but leave ratings to specialists or operating people.

In physics, Heisenberg's uncertainty principle states that both the position and state of certain sub-atomic particles cannot be known. This is because the process of measurement distorts either the position or state. A parallel phenomenon occurs in space planning. The process of judging and evaluation often leads to other options. Thus, some or all of the space plans may change as a result of the evaluation process. Or, a hybrid space plan that features the best parts of several original options may emerge.

Block 5 of the procedure diagram examines the decision criteria and available tools. Two to four tools appropriate for the evaluation should be selected.

Block 6 evaluates all options with respect to the identified qualifiers. Any option that fails to meet a qualifying criterion drops from consideration.

Block 7 evaluates the options with respect to the decider criteria. New or hybrid options go on the list of available options.

After evaluation, one option is selected for development. A decision summary recapping the decision process should be prepared. The summary and decision make up Block 10.

Evaluating the Cosmos space plans

The Cosmos design team and steering committee met to evaluate the proposed space plans. They decided that both the steering committee and design team should participate in the evaluation. They first reviewed the original project objectives. These came from Task 03.01, "Plan Project":

- reduce material handling cost;
- reduce operating costs;
- deliver project under budget of $800,000;
- improve delivery performance;
- improve teamwork, communication, and quality;
- allow for new products; and
- accommodate 1998 production.

From the original objectives, they derived these decision criteria:

D	Material flow
D	Direct operating cost
Q/D	Initial cost
D	Delivery
D	Communication
D	Teamwork
D	Quality
D	New product adaptability
Q	Meets 1998 production requirement
Q	OSHA/EPA Compliance

A "Q" notation designates the qualifiers. OSHA/EPA compliance is necessary for any space plan. Those that fail to meet this qualification are no longer considered. Similarly, the 1998 production requirement is a qualifier. Initial cost is both a decider and qualifier. A space plan must meet the $800,000 budget limitation to be considered; this is the qualification. Initial cost below $800,000 is a benefit; this is the decider. All the other criteria are deciders, denoted by a "D."

The Cosmos team chose PNI, MFA, cost estimating, payback, and weighted factor analysis as the tools for evaluation.

They analyzed material flow first and then used the results to assist with the cost estimating. In Step 1 of the MFA, they developed the flow diagrams illustrated in figure 3.40.

These diagrams show where the flow complexity for the existing

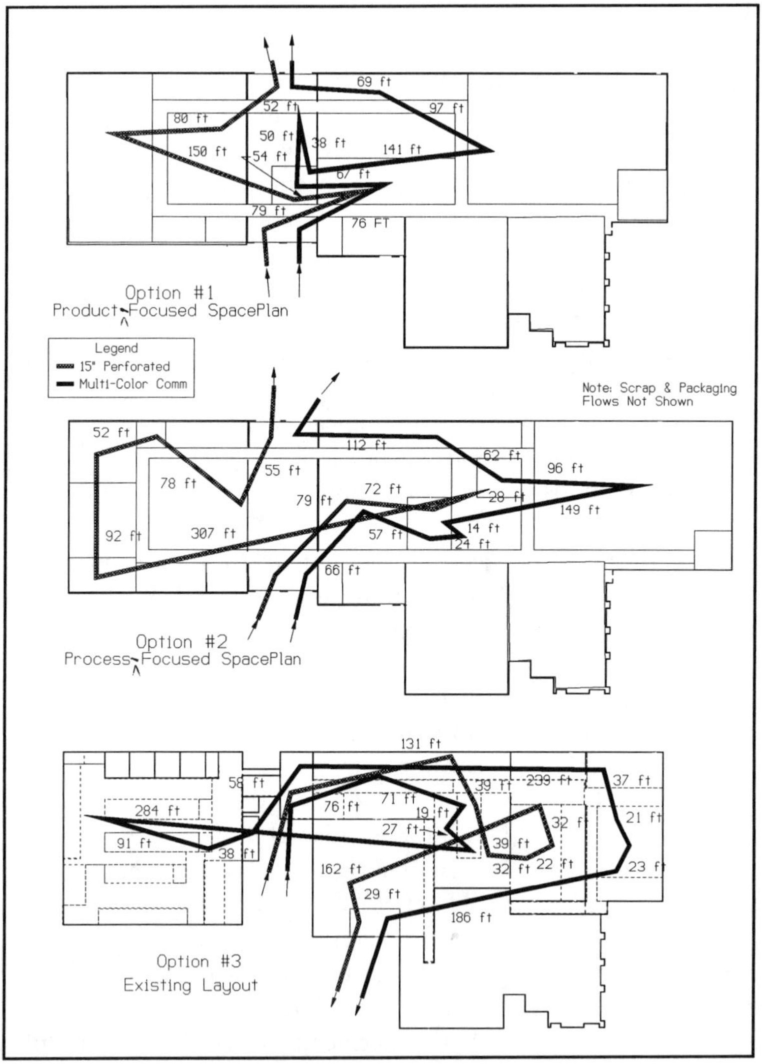

Figure 3.40 - Material Flow Evaluation

layout is greatest. Option 2, a revised functional layout, improves the flow complexity and shortens flow distance. Option 1 improves complexity and further shortens the total distance. The flow complexity index (FCI) counts the frequency of flow crossings on the diagram. Option 1 has an index of 0, Option 2 has an FCI of 4, and Option 3 has an FCI of 6. Visual examination of these material flow charts confirms the increasing material flow complexity from Option 1 through Option 3.

Transport work is the summation of each flow distance multiplied by the flow rate. The units for Cosmos are EFUs per day. Table 3.4 is the spreadsheet used by the team to calculate distance and transport

Option #1

(A) From -To	(B) From	(C) To	(D) Units	(E) Fwd Units /Day	(F) Rev Units /Day	(G) EFU Fact	(H) Tot EFUs /Day	(I) Dist Ft	(J) Trans Work
01-04	Silkscrn	Post	Scrn	SHTS	69.0	0.17	12	97	1118
01-05	Silkscrn	Slit	SHTS	69.0	0.17	12	141	1625	
03-05	Roll	Ops	Slit	SLITS	46.0	0.11	5	150	759
03-08	Roll	Ops	FG Stge	PKGS	2134.0	0.01	19	80	1536
03-10	Ops	Receiving	PMAT	7.3	0.25	2	156	285	
04-09	Post	Scrn Shipping	SHTS	69.0	0.25	17	69	1190	
05-06	Slit	Int Stge	SLITS	46.0	46.0	0.11	10	43	435
05-07	Slit	RM Stge	LOGS	23.4	0.50	12	67	784	
05-09	Slit	Shipping	LOGS	0.2	0.50	0	93	9	
07-10	RM	Stge	Receiving	LOGS	23.4	0.50	12	78	907
08-09	FG	Stge	Shipping	PKGS	2134.0	0.01	19	52	999
							119	**1026**	**9647**

TABLE 3.4

work for Option 1. The other options have similar spreadsheets. The three options have transport work of 9,647, 18,669, and 28,131 EFU-feet per day respectively.

Another measure of material flow is the frequency count for material moves. Option 1 has 11 internal moves, Option 2 has 14, and Option 3 has 21. The total distance traveled for the two representative products is another measure. Options 1 through 3 have distances of 1,026, 1,723, and 2,135 feet, respectively. The average number of trips per day is 119, 132, and 198.

This analysis assumes that all trips use the EFU, an equivalent handcart, as the means. When implemented, the layout actually will use several methods of handling. However, for estimating, the EFU assumption is a reasonable approximation.

Figure 3.41 is a graphic display of the MFA results. Based on every material flow measure, Option 1 is significantly better than Option 2. Option 2 is significantly better than Option 3.

Financial analysis

Table 3.5 summarizes the financial results for the three options. Option 3—the existing layout—maintains the status quo. For this reason, there is no change in either savings or costs. Option 3 thus provides the baseline for the financial analysis.

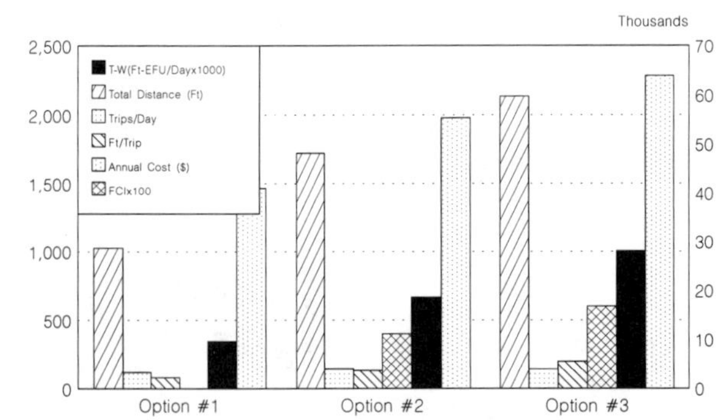

	Option #1	Option #2	Option #3
T-W(Ft-EFU/Dayx1000)	9.647	18.669	28.131
Total Distance (Ft)	1,026	1,723	2,135
Trips/Day	119	142	142
Ft/Trip	81	132	198
Annual Cost ($)	41.071	55.399	63.957
FCIx100	0	400	600

Figure 3.41 - Material Handling Summary

The center building is the new construction between the two existing buildings. The team estimated the cost at $35 per square foot. The east extension for Option 2 will cost about $30 per square foot because it does not have loading docks.

Option 2 will need new equipment, valued at about $23,000, which will cost $21,000 for installation. Option 3 requires more equipment because of its cellular nature. Rearrangement costs are $45,000 and $28,000, respectively, for Options 1 and 2.

The cellular approach of Option 1 will require significant training and additional consulting fees when compared to Option 2. The team also anticipated a more difficult start-up.

A contingency of 15 percent that allows for unplanned costs is applied to the implementation of both new options. Either Option 1 or

Initial Cash Outflows Description	Option 1 Amount	Option 2 Amount	Option 3 Amount
Center Building	$161,000	$161,000	$0
East Extension	$0	$307,800	$0
Equipment	$176,000	$23,000	$0
Installation	$49,500	$7,800	$0
Rearrangement	$45,000	$28,000	$0
Training	$32,000	$0	$0
Consulting	$43,500	$20,000	$0
Startup	$100,000	$45,000	$0
Contingencies	$91,050	$88,890	$0
Total	$698,050	$681,490	$0

Annual Inflows Description	Option 1 Amount	Option 2 Amount	Option 3 Amount
Increased Sales	$750,000	$750,000	$0
Material Handling	$22,885	$8,558	$0
Direct Labor	$132,000	$10,000	$0
Other Indirect	$75,000	$37,500	$0
Working Capital	$140,000	$0	$0
Quality	$230,000	$20,000	$0
Total	$1,349,885	$826,058	$0
Initial Inflow Inventory	$1,750,000	$0	$0
Years-To-Payout	0.23	0.82	n/a

TABLE 3.5

Option 2 will bring increased sales and production. The net profit for this is estimated at $750,000.

Material handling savings come from the decrease in handling and transport work. Using the data from the material flow analysis, the team estimated cost savings of $22,885 per year for Option 1 and $8,558 dollars per year for Option 2. This assumes four minutes of loading and unloading for each trip. It assumes an average transport speed of 150 feet per minute and an $18.50 hourly labor cost. It also assumes that each move has an empty return trip.

Calculations for direct labor, quality, and other indirect labor savings are less rigorous, but the team developed conservative estimates from their experiences.

Option 1 has a significant inventory reduction of $1.75 million. This is a one-time savings and lessens the working capital required. The interest on this, at 8 percent, amounts to $140,000 per year.

The payout for Option 1 is 0.23 years. The payout for option 2 is 0.82 years. Both payouts are quick. There is no payout for Option 3 because there is no initial investment.

An ROI analysis would be more rigorous than the payout method. However, the fast paybacks for Options 1 and 2 indicate that the increased complexity and effort required for an ROI analysis is unnecessary.

The team reviewed the decision criteria to see if the options met all qualifiers. All three options met the regulatory qualifiers. All three options met the budgetary qualifier. Only Options 1 and 2 will satisfy 1998 production requirements. This signifies that doing nothing, Option 3, is not a viable course of action. Option 3, however, has been useful as a baseline for improvement estimates.

Positive-negative-interesting

Having completed the quantitative analyses, the evaluation group then turned to PNI analysis. Meeting with a facilitator, they focused on each option and each aspect in turn. They used brainstorming techniques to develop the positive, negative, and interesting points for each option. The results are in table 3.6.

Weighted factor analysis

With the quantitative and qualitative analysis complete, the team turned to weighted factor analysis, where analysis and opinion are merged into a single decision.

Figure 3.42 summarizes the weighted factor results. The team first reviewed each factor and confirmed the definition. Through discussion, the members reached a consensus on the weights. Each factor had a weight between one and ten.

Operating cost, quality, and delivery received high weights (ten and nine). These factors have the most direct effect in the marketplace. The group believed they had the highest strategic importance.

Material flow, communication, and teamwork received weights in the seven to eight range. These factors are somewhat related. Good material

PNI Analysis Summary

Option 1 Positive	Option 2 Negative	Option 3 Interesting
Best Material Handling	High Training Required	
Simplification	It Might Not Work	
Neat & Clean Geometry	Difficult Adjustment	
Less Inventory	Higher Risk	
Better Teamwork		
Fits W/TQM		
Faster Throughput		
Faster Response		
Less Space		
Less Cost		
Faster Payout		
Best Annual Cost		
Easier Supervision		
Employee Involvement		
Nice Aisle System		
Uses Current Software w/Kanban Production Control		

Option 1 Positive	Option 2 Negative	Option 3 Interesting
Reduced Material Handling	Low Risk	Allows Cellular Transition Later
Nice Aisle System	High Cost	
Neat & Clean Geometry	Allows Cellular Procrastination	
Easy Personal Adjustment	No Throughput Improvement	
Lots of Space	More Space Required	
	Lower Payout	
	Higher Annual Cost	
	Less Employee Involvement	
	Does Not Assit TQM	
	Less Teamwork	

TABLE 3.6

flow improves communication and simplifies operations. It also reduces interdepartmental problems. Better flow and better communication enhance teamwork. Material flow also ties to operating cost.

The team then examined material flow and compared the three options. With quantitative data from the MFA, they quickly achieved consensus on the ratings. Option 1 received an A for material flow,

Weighted Factor Analysis

Project: FACILITY RE-ENGINEERING		Prj #: IE01
Company: COSMOS PRODUCTS	By: QL	Date: 12/12/94
Note: COMBINED SCORES		

ID	Factor	Wght	Option #1		Option #2		Option #3		Option #4		Option #5		Notes
			Rate	Score	Rate	Score	Rate	Score	Rate	Score	Rate	Score	
F1	MATERIAL FLOW	8	A	32	I	24	U	0					
F2	COMMUNICATION	7	I	14	O	7	U	0					
F3	TEAMWORK	7	A	28	U	0	U	0					
F3	NEW PRODUCTS	2	O	2	E	6	E	6					
F4	INITIAL COST	3	I	12	E	9	A	12					
F5	DIRECT OPER COST	10	E	30	I	20	U	0					
F6	QUALITY	9	A	36	I	18	U	0					
F7	DELIVERY	10	A	40	O	10	U	0					
		Totals	194		94		18						

#	Option Description
1	CELLULAR
2	FUNCTIONAL
3	EXISTING
4	
5	

Notes:
JUDGES:
O. W. HOLMES F. TAYLOR
J. MARSHALL H. FORD
W. BURGER
E. WARREN

Figure 3.42 - Weighted Factor Analysis

Option 2 received an I, and Option 3, the baseline, received a U. The team repeated this process for the other factors.

Delivery had some quantitative basis. The number and length of moves from the MFA indicated that delivery would improve under Option 3, whereas delivery perhaps would see slight improvement under Option 2.

Quality, new products, teamwork, and communication had no quantitative analysis. Nevertheless, discussion and a review of the PNI analysis brought a consensus among the team members.

They then multiplied each factor weight by each rating and totaled the score for each option. Option 1 received 194 points, Option 2 received 94 points, and Option 3 received 18 points. From those scores, the group concluded that Option 1 was best by a large margin.

The team could have begun the weighted factor analysis by weighing each factor individually and rating each option. They could then average and compare their results as a basis for discussion. This is a useful technique when it appears that individuals have widely differing views. Several computer programs are available for this type of multifactor decision making, but the most important results derive from the discussions. In most situations, a manual compilation like that in figure 3.42 is sufficient.

Conclusion

This completes the discussion of macro-space-planning. Many of the methods apply to other levels of space plan design. Material flow analysis, for example, is an important tool for Level 2, "Site Planning." Weighted factor analysis applies at all levels.

For most facility planning, the macro-space-plan is the most important planning level. It is where strategy is defined and the first steps toward implementation are taken. It is the level that usually has the greatest impact on a firm's competitive position. For these reasons it has been the subject of the most substantial discussion in this book.

Chapter 4

Micro-Space-Planning and Work Cell Design

Micro-space-planning, Level 4, is the next level of design detail. Micro-space-planning determines the location of equipment, furniture, and workstations for each department or SPU of the macro-space-plan. Process charts and operation plans define how people and equipment will work within the space (fig. 4.1). The output is usually a two-dimensional layout for each department or SPU. It may be helpful to combine the space plans for each SPU on a large drawing that overlays the macro-space-plan.

Sound micro-level space planning ensures that work flows smoothly in each SPU, promoting teamwork. It affects supervision, organizational learning, costs, inventory, quality, delivery, flexibility, and coordination, as well as many aspects of the firm's cost structure. It provides an opportunity to reengineer the processes.

During this activity, collaboration with supervisors and workers in each department is vital. The success of the space plan depends largely on their cooperation because it is their space and their work that will change. Upper management usually participates less at this level than at higher levels.

Micro-space-planning sometimes occurs before the large-scale macro-space-plan is designed, particularly if only one area needs a new space plan. Or, a designer may wish to develop an experimental work cell to gain experience and try out new ideas. If so, he or she should review the parts of Chapter 2 that discuss levels and phasing.

The approach to micro-layouts discussed in this chapter is similar to that for macro-layouts. It has five major tasks, each of which has a corresponding procedure diagram that tells the designer or design team how to do it. The macro-space-plan's fundamental elements—SPUs, affinities, space, and constraints—still apply, yet there are significant differences. This chapter explains the tasks and procedures.

The model project plan helps designers organize this work, while allowing them to look ahead and behind. Focusing on a detailed task also means stepping back to view the entire work cell design. This differs significantly from designing a macro-space-plan, where each task can be approached as a relatively independent piece of work.

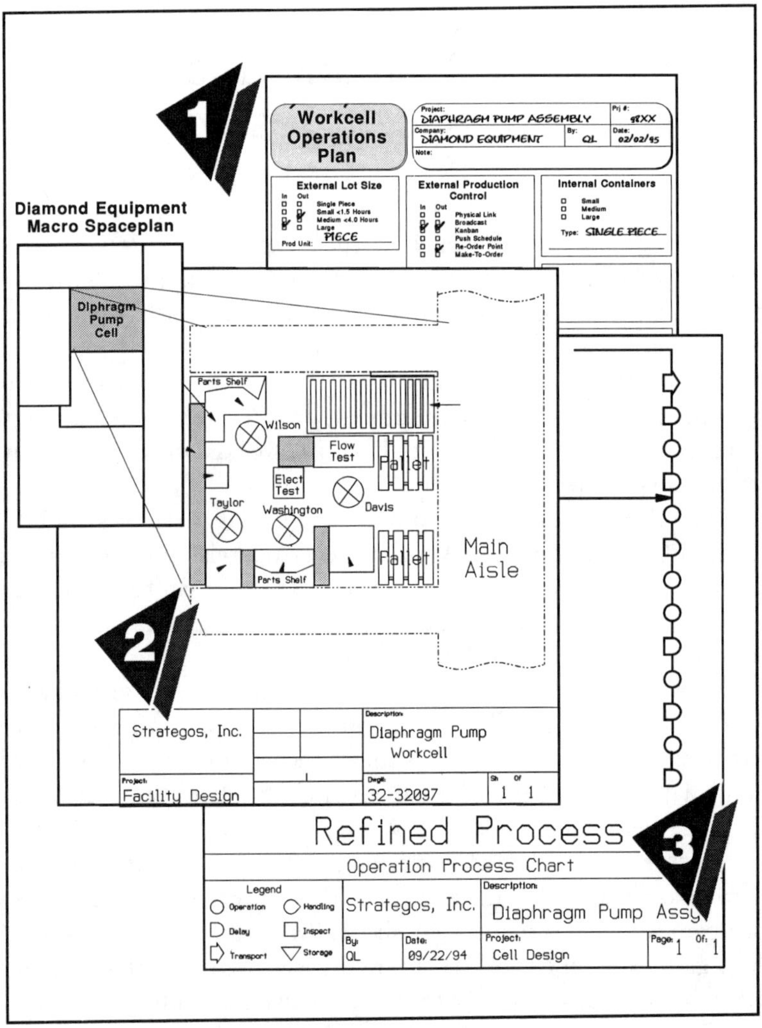

Figure 4.1 - Workcell Operations Plan

Work cells

Macro-space-plan SPUs are often work cells. Work cells are small, self-contained work units with several machines or operations. Their equipment and people are situated together in a compact, sequential arrangement. Work cells usually have two to ten people and two to ten operations. They perform all or most operations necessary to produce their products or complete a major production sequence. Here are some examples:

A manufacturer of jet-engine turbine blades decided to use work cells to prepare the intricate and delicate wax patterns for precision investment castings. A molding machine produced wax components. A team of two to five people then assembled the components into patterns. They manually cleaned edges, filled voids, and inspected completed patterns. Within a few days of cell startup, defects and rework declined 30 percent while productivity rose significantly. Throughput went from several weeks to one day.

The same manufacturer of turbine blades also chose to use cells for processing raw castings. In the post-casting cell, two to seven people cut gates and risers. They shot-blasted the blades, ground flash, de-burred, x-rayed and inspected. Throughput time went from months to days. Quality increased dramatically. Productivity improved.

A maker of industrial air cleaners decided to use cells for final assembly. Productivity increased by 20 percent. Response time for special orders went from weeks to hours. Finished-goods inventory declined by 96 percent. The same manufacturer then linked a weldment cell for cabinets to the final assembly cells. This further reduced WIP inventory and improved quality. The air cleaner manufacturer then placed sheet metal fabrication in a linked cell. Internal-parts shortages disappeared. WIP declined further.

A firm that manufactures mechanical control cables chose to use cells for final assembly. That company reduced the amount of finished goods by 90 percent and improved quality by 60 percent. Workers in the cell consistently produced at 120 percent of the former work standard. In addition, workers in the cell performed many tasks that indirect employees previously accomplished.

A manufacturer of HVAC equipment machines and assembles large gearboxes. The firm installed a machining cell and a linked assembly cell and reduced finished goods by 80 percent while improving quality and productivity significantly. Response time for special orders went from weeks to twenty-four hours.

Many other industries use work cells with similar results. These include those specializing in electronics, steel casting, structural fabrication, plastics, and vinyl windows. Cells work with single products or many products, in offices, hospitals, and service industries. They are effective wherever coordination of sequential operations dominates a process.

Cells can be classified by their range of products and processes (fig. 4.2), using the concept of operation focus discussed in Chapter 3 (see figs. 3.19 and 3.20). These cell classifications are:

Dedicated Cell—A work cell that produces only a single product with, perhaps, minor variation. It has multiple, sequential processes. Dedicated cells have a high degree of product focus. Such a cell may be found in the lower left region of the chart in figure 4.2.

Group Technology (GT) Cell—GT cells produce a family of related products. They have similar, but not necessarily identical, processes. Examples are the machining of similar shafts or the assembly of products with many common parts. GT cells have a moderate degree of product focus.

Functional Cell—These cells use a single process, such as heat treat, that operates on a wide variety of unrelated products. Functional cells are process focused. They may be found on the left in figure 4.2.

Project Cell—These cells produce a wide variety of unrelated products using multiple processes. A small tool-and-die shop is, effectively, a project work cell. Project cells have neither product nor process focus.

Project and functional cells are outside the common notion of cellular operations. Nevertheless, they can achieve some of the advantages of traditional cells. They use the same design approach.

Interpreting terminology and conventions

Most business operations provide a variety of products and services. Some products are for external customers while others go to internal customers. Although services are intangible, their provision follows the same principles as a manufactured product. In this chapter, for space plan purposes, the term product may also be interpreted as service.

Also, the sizes of lots and containers in work cells have different interpretations in different industries. For example, the term large container means one thing to a person who works in an electronics plant, but it has quite a different meaning to someone with a heavy steel background. It may have no meaning at all to someone in the insurance industry. Conventions are used as approximate, but well-understood, terms that cross industry boundaries.

The convention used in the examples in this book are based on design production volume. *Single piece* refers to one item or a single unit of production. *Small* is a quantity of product equivalent to less than one half-hour of production. *Medium* is a quantity of product equal to or less than four hours of production. *Large* is a quantity of product greater than four hours of production.

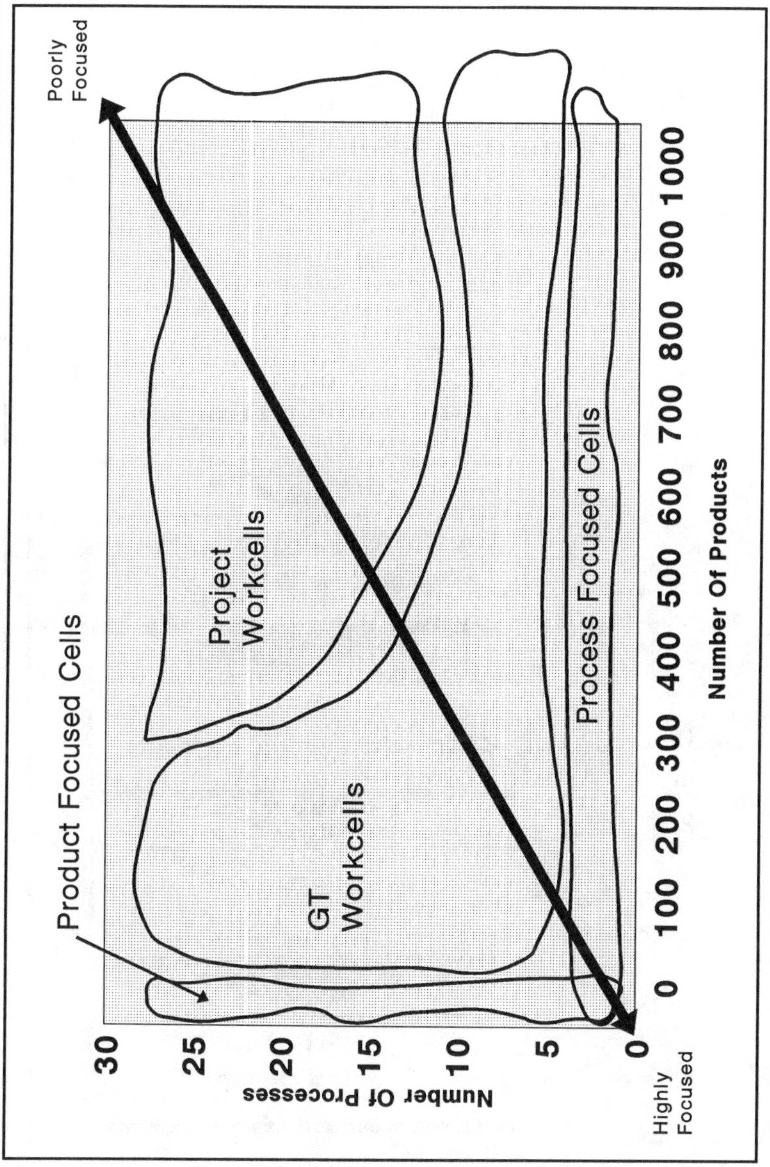

Figure 4.2 - Workcell Focus, The Product-Process Plane

The micro-level tasks

The preferred approach to micro-layouts follows the pattern developed in Chapter 3 for macro-layouts. Figure 4.3 is the model project. It has five generally sequential tasks:

- Task 04.01, "Analyze and Select Products";
- Task 04.02, "Design and Refine Process";
- Task 04.03, "Select Infrastructure Elements";
- Task 04.04, "Design Cell Layouts"; and
- Task 04.05, "Select Cell Option."

Each task has a procedure diagram that shows the design steps and their sequences. Various forms and aids help make design decisions. However, there are parts of each task that can have multiple options and many

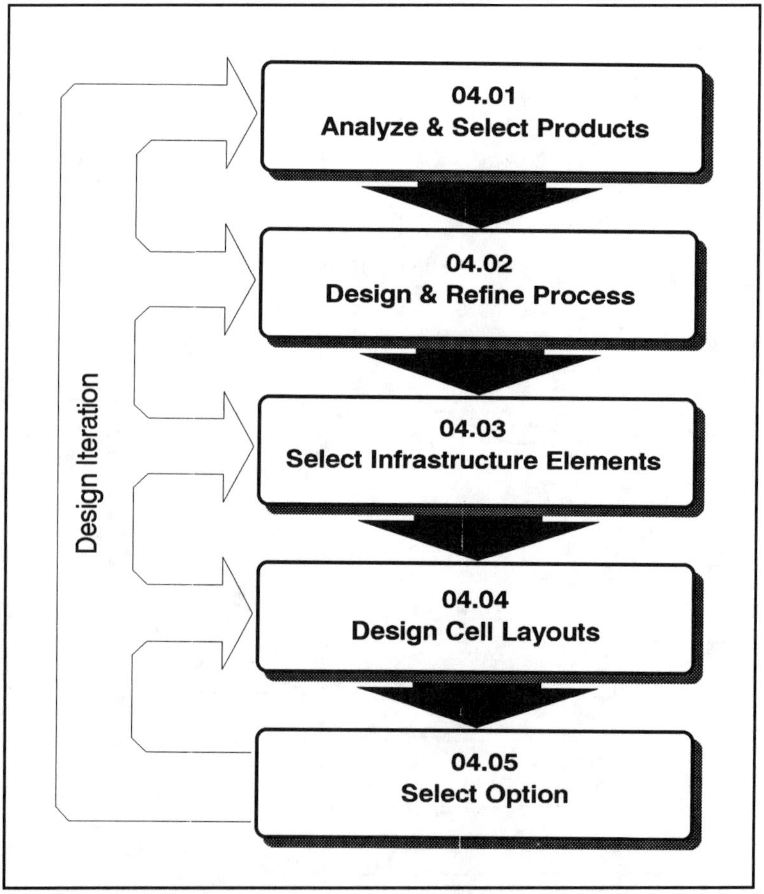

Figure 4.3 - Micro-Layout Model Project Plan

points where a particular line of inquiry appears infeasible. Multiple design iterations may be necessary.

For example, the selection of a particular product group appears viable during Task 04.01, but the calculation of process time during Task 04.02 leads to the realization that the equipment has very low utilization. A return to Task 04.01 for a product group revision may be necessary. Or, during the evaluation process, it may become apparent that a different option for production control offers certain advantages. Production control is an infrastructure element from Task 04.03. The designer then returns to that task, specifies a new option, and designs an additional layout.

Some micro-layouts fall outside the work cell classifications mentioned earlier. They use SPUs and departments with many more people and processes. Such operations are more difficult to coordinate, control, and supervise. They are often process-focused. Non-cellular micro-space-plans require less rigor and effort in Tasks 04.01 through 04.03. Otherwise, the same model project and task procedures apply.

Compatibility of design elements

Several blocks in the task procedures include the question, "Compatibility OK?" This is a signal to refer to a compatibility guide. Figure 4.4 is a compatibility guide for the elements of a work cell design and should be read in the same way as the affinity chart. Shaded bullets indicate usual compatibility between two elements. The example in the lower right shows that a small external lot size is compatible with a small external container and incompatible with medium and large external containers.

This guide should be consulted frequently for the first few work cell designs, although experienced designers may not need to use it. It is not infallible, however. There are occasions when a design will use elements that, under normal circumstances, are incompatible.

Selecting the products

Product selection is the first and most fundamental task for work cell design. It must be decided which products each SPU will process. This determines the focus for the cell. All other design decisions flow from this task. The deliverable for this task is a list of products for each cell and a design production volume for each product.

Decisions are based partially on how the products are processed. A functional cell might process every product. In a predominantly functional space plan, a product may travel through many SPUs. In a product-focused work cell, the products may stay in one cell.

During the macro-space-planning, products may have been

assigned to a composite cell or designers may have made a selection, knowing that further refinement would be necessary later. Task 04.01 is the time when this should occur.

Product selection groups products and processes. In most situations, to allow product focus, the groups should consist of similar products that use a similar process and sequence.

Figure 4.5 shows the procedure for Task 04.01, "Select Products." Block 1 represents input data for products and volumes. This data may have been collected during Task 03.02 of a previously prepared macro-

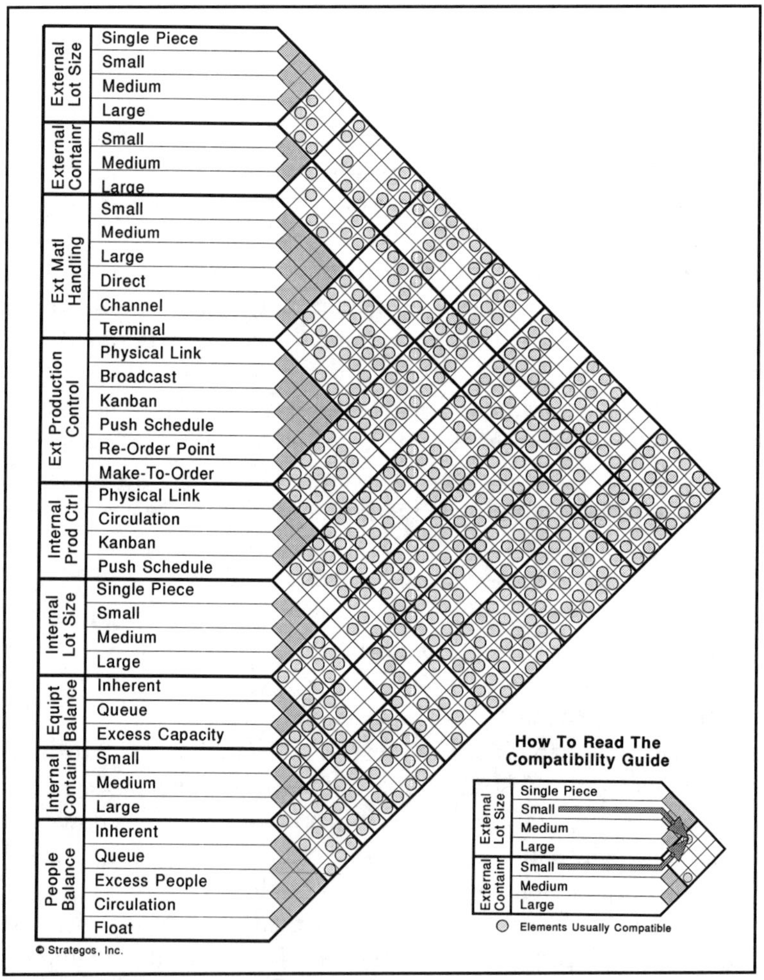

Figure 4.4 - Workcell Compatability Guide

space-plan, but more detail may be needed at this point. Block 2 represents input data in the form of information about the current or proposed processes. Again, some of this may have been collected for the macro-space-plan, but more detail may be needed at the micro-level.

In Block 3, a product profile and sales forecast is prepared for individual products rather than product groups. However, if the variation between products is minor, it is possible to consider them a single product.

Blocks 4 through 9 help select the methodology for product grouping and selection. The basis for this grouping varies. In machining

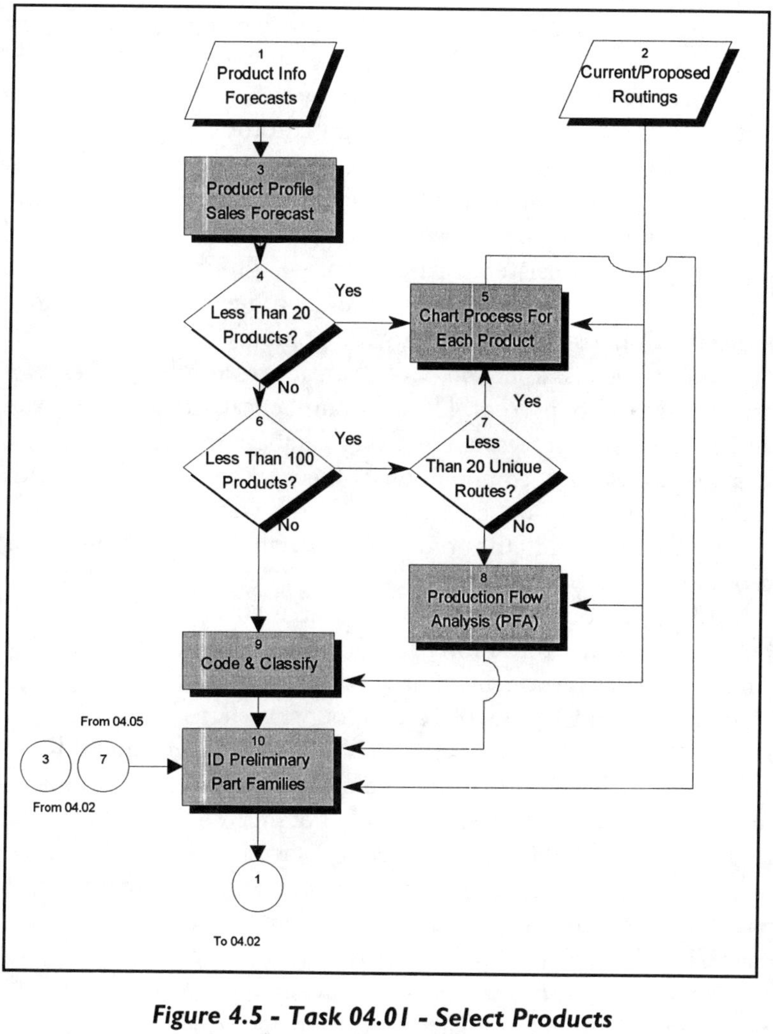

Figure 4.5 - Task 04.01 - Select Products

operations, similar size and features such as round shafts or milled shapes are most likely the determining factors. In assembly operations, part commonality is the criterion.

Three methods exist to help with this selection: intuitive charting/grouping, production flow analysis, and coding and classification. In intuitive grouping, an experienced person or team examines the products, often visually and perhaps with process charts to bring out valuable information. From the results, that person or team assigns products to logical groups. This is a quick method that works well if there are twenty or fewer products or in instances where a few products can represent a larger population.

Experienced shop people often think they can group many parts with complex routings, using only their experience and intuition. This becomes less and less effective beyond twenty distinct products. Instead, intuitive grouping should be used to design a prototype cell for learning.

Product flow analysis (PFA) uses current routings and processes. With spreadsheets, computer analysis, or manual methods, parts that have similar routings are grouped. PFA is a powerful and fast method if processes are consistent. Like intuitive grouping, it is most advantageous when processes are simple or products few. PFA begins to lose effectiveness with more than 100 products.

Coding and classification (C&C) associates a code number with every product and/or process. The code number carries information on product features or process methods. A database manager then sorts and retrieves various combinations of codes to group the items appropriately. Coding and classification works well even when tens of thousands of items are in the product mix. It handles complex products and processes.

Although C&C requires a major initial effort, the benefits are commensurate with the effort. It is a valuable method because it examines the database from many perspectives. Families can be identified for purchasing and the issuance of group purchase orders. Design features can be examined and parts, tooling, and gauges standardized. Cost drivers can also be identified and costs allocated more rationally.

Coding systems have been the subject of much research during the past thirty years. The advent of modern database managers has made coding systems flexible and easier to use. Nevertheless, there is a significant element of art to their use. Coding systems can be quite sophisticated. Designers and planners should consult references such as Salvendy's *Handbook Of Industrial Engineering* before investing a large effort in a C&C project. Experience is a major factor in getting good results.

Block 4 asks if there are fewer than twenty products. If so, Block

5 recommends preparing process charts for each product and then proceeding to Block 10.

If there are more than twenty products, but fewer than 100 products and fewer than twenty unique routings, a process chart for each routing is necessary. For twenty or more routings, a production flow analysis is recommended. If there are more than 100 products, they should be coded and classified.

The break points of twenty and 100 products are approximate. When processes are simple or products are similar, charting or PFA can be used for even very large mixes.

In Block 10, preliminary part families are identified based on the previous analysis. For assembly operations, these are products that use the same parts. For machining, or other equipment-intensive operations, these are parts that use the same equipment and tooling. For services, these are products that have a common customer base or use similar skill sets.

At this point, it has not been determined whether the families of equipment and people will form a viable cell. A preliminary product family might have too many or two few people to create an effective cell. The family might, or it might not, utilize the available equipment effectively. Proceeding through the analysis will test the viability of the current family selection. If the family does not appear viable at any time, options include: adding more products, splitting the family, or modifying the process.

The Diamond Equipment Company

Throughout this chapter, the activities of the Diamond Equipment Company will be used as examples of how micro-space-planning tasks and procedures may be accomplished.

Diamond manufactures a variety of special pumps for chemicals and hydrocarbons and is reengineering their facilities. During the macro-space-planning, the staff identified fourteen product groups for cellular manufacturing using the intuitive method. Each product group has a corresponding work cell for assembly.

One of Diamond's proposed work cells assembles diaphragm pumps. These products appeared to be a good selection for a work cell because they used many common parts and the same test equipment. Their design characteristics were similar. The total assembly and indirect labor originally used five to six persons.

The "A Team" was chosen to design this cell as its first project. The team members included a supervisor, five workers from various production departments, and an industrial engineer experienced in space planning systems.

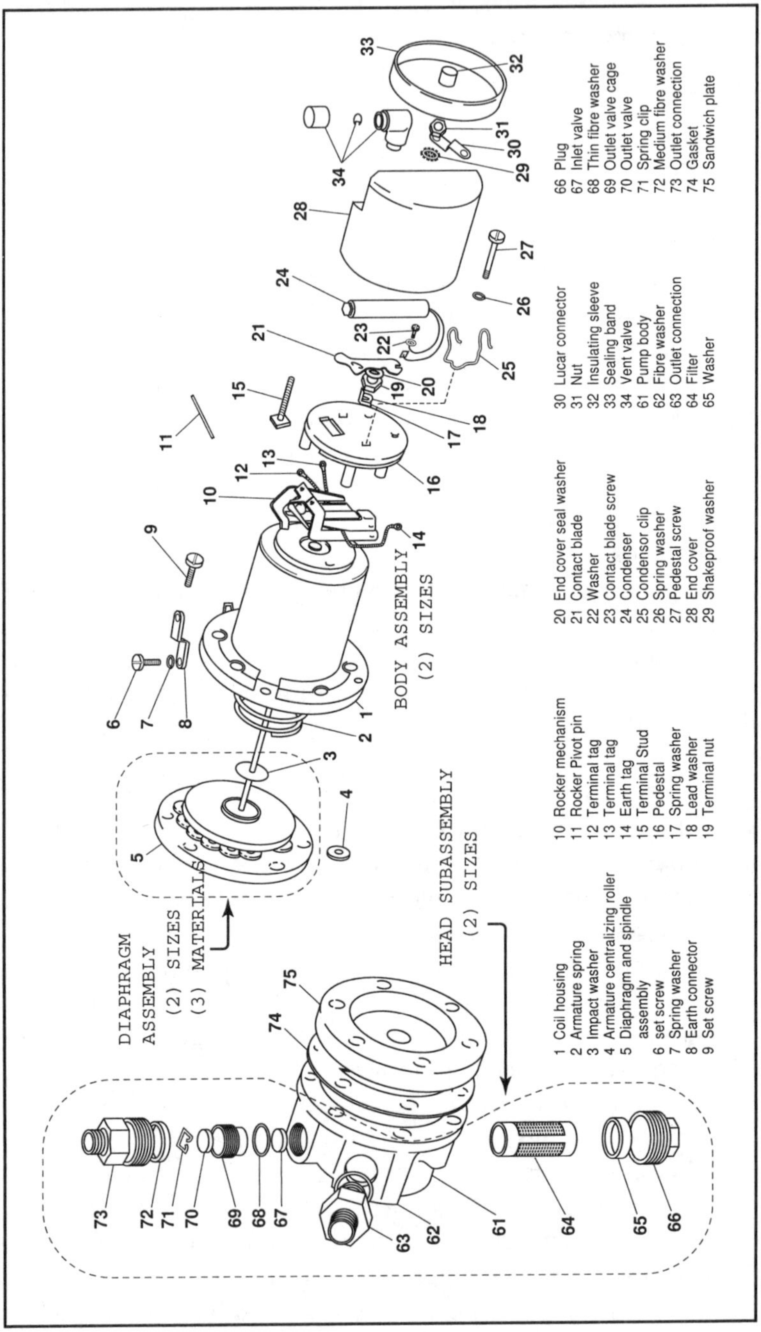

Figure 4.6 - Drawing of a typical product

The initial product selection was still tentative. The A Team's first task was to verify that selection. Team members had the option of dividing the diaphragm pump cell into several cells. If necessary, they could add other pumps to the cell's product family.

The team prepared a P-V analysis, including a drawing (fig. 4.6) of one typical product. From the drawings and other information, the team learned that their proposed product mix included two pump sizes. They also discovered that three pressure ratings for each size required six different springs and three diaphragm materials required six diaphragms. Electrical components for each pump were the same.

A forecast for all proposed products is shown in figure 4.7. The team chose a baseline design volume for the work cell of eighty-eight units per day. The product profile in figure 4.8 shows relative volumes based on recent history. The team decided that this product selection was viable and proceeded to Task 04.02, "Design The Process."

Diamond also has a machine shop with twenty-seven people and a variety of numerical control (NC) and conventional equipment. The machine shop uses vertical and horizontal machining centers, lathes, drills, and secondary equipment. The machine shop makes components for all assembly operations.

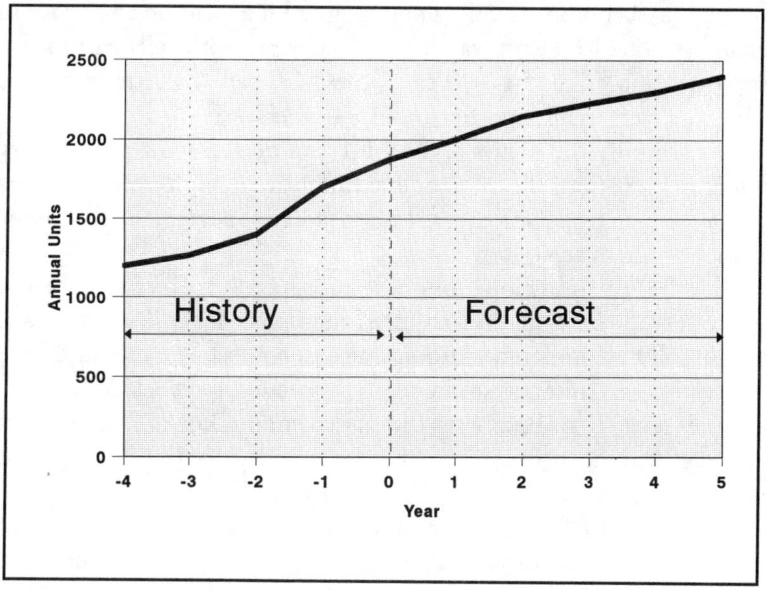

Figure 4.7 - Sales History/Forecast For Diaphragm Pump Workcell

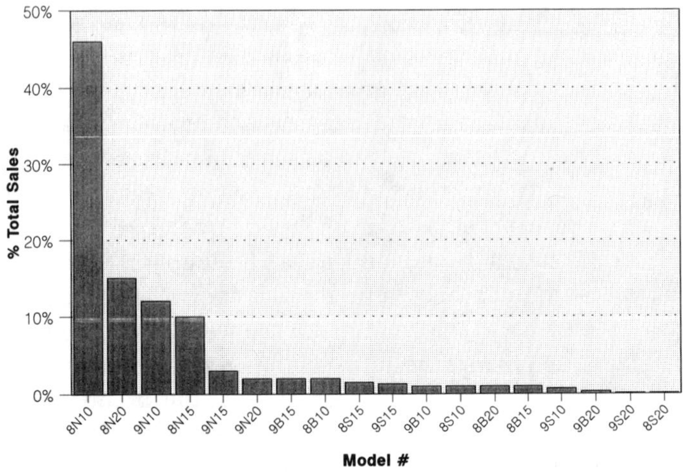

Figure 4.8 - Product Profile for Diaphragm Pump Workcell

At the macro-space-plan level, the designers identified the machine shop SPU as a composite cell. They assigned it a location and space in the facility. The M Team was chosen to design work cells within this composite cell. Like the A Team, the M Team had an industrial engineer, a supervisor, and five other members.

To simplify the analysis, the M Team performed a preliminary product segregation. Team members identified high-volume items that ran on dedicated equipment. They decided to segregate these items and analyze them separately.

For the remaining low-volume items, the team utilized figure 4.5. They found the information required in Blocks 1 and 2 of this task procedure. They prepared a product profile and sales forecast. There were twenty-seven items, so the team proceeded to Block 6. Because there were fewer than 100 items, the team went on Block 7. They found twenty-six unique routings among their twenty-seven items. The procedure suggested production flow analysis.

To apply PFA, the team developed a matrix (fig. 4.9) that correlated parts and processes but did not show the sequence of operations. Part identifiers were listed on the left, equipment or work centers across the top. The check marks in figure 4.9 show that a particular part uses a particular work center.

	Manual Drill & Tap	C'less Grinder	12" Engine Lathe	HiT30S CNC Lathe	40x20x25 VMC	NC Drill	W&S #3 Turret Lathe	15x10x12 VMC	Slotter	Heat Treat	Snag Grinder	B&S #2 Scrw Mach	Manl Broach	Horiz Mill	Manl Band Saw	HT-8G CNC Lathe
32-96478 Pin	✔	✔										✔				
34-65156 Base						✔		✔								
34-85427 Impellor				✔									✔			
34-87654 Impellor				✔							✔					
32-56261 Shaft	✔						✔							✔		
32-56781 Shaft	✔	✔					✔							✔		
34-65427 Impellor				✔												
32-91789 Shaft		✔	✔							✔				✔		
34-92461 Cover				✔		✔					✔					
34-92954 Pump Base								✔			✔					
34-65786 Base						✔		✔			✔					
32-86454 Plunger	✔	✔								✔		✔				
34-81454 Impellor				✔							✔		✔			
32-01768 Shaft					✔					✔					✔	
32-45789 Shaft	✔	✔	✔							✔				✔		
34-92941 Frt Cover						✔					✔					
32-34654 Shaft		✔								✔				✔	✔	✔
34-14267 Cover				✔		✔					✔		✔			
32-92063 Clamp Screw		✔							✔			✔				
32-98763 Spec Screw									✔			✔				
32-96768 Baseplate								✔					✔			
32-39814 Pump Shaft										✔					✔	✔
32-39768 Shaft					✔					✔					✔	✔
34-14236 Rear Cover				✔		✔					✔					
32-89814 Plunger		✔								✔		✔	✔			
32-96742 Base								✔								
32-45478 Center Pin	✔											✔				

Figure 4.9 - Production Flow Analysis

Initially, the matrix appeared to have no identifiable pattern. The prospects for finding viable work cells and part families did not seem good. The team examined the matrix further. They sorted and rearranged rows and columns. Gradually, structure emerged from apparent chaos; the result is illustrated in figure 4.10. The shaded areas represent machine groups and part families for four work cells.

The upper left block is a manual machining cell with turning, milling, and other equipment. The team considered breaking this cell into turning and milling cells. However, the production volume seemed insufficient. Moreover, the cell had a compatible set of skills.

The next block is an NC cell. It turns and mills shafts. This cell shares a heat treat furnace with the manual cell. The third cell mills and drills bases and plates, usually castings or weldments.

A fourth cell turns, mills, and drills large rotational parts—mostly castings—and shares an NC drill with the third cell.

Two operations fall outside the cells. These use a snag grinder and manual broach, which are low-cost machines. The team expects to purchase additional such machines. Every cell that requires these operations will have a snag grinder and broach, even though they will sit idle much of the time.

	Slotter	B&S #2 Scrw Mach	12" Engine Lathe	Horiz Mill	C'less Grinder	Manual Drill & Tap	W&S #3 Turret Lathe	Heat Treat	HT-8G CNC Lathe	40x20x25 VMC	Manl Band Saw	15x10x12 VMC	NC Drill	HiT30S CNC Lathe	Manl Broach	Snag Grinder
32-96478 Pin		✔			✔	✔										
32-45478 Center Pin		✔				✔										
32-98763 Spec Screw	✔	✔														
32-92063 Clamp Screw	✔	✔			✔											
32-56781 Shaft				✔	✔	✔	✔									
32-56261 Shaft				✔		✔	✔									
32-91789 Shaft			✔	✔	✔			✔								
32-45789 Shaft			✔	✔	✔	✔		✔								
32-86454 Plunger		✔			✔	✔		✔								
32-89814 Plunger		✔			✔			✔							✔	
32-34654 Shaft				✔	✔			✔	✔		✔					
32-39814 Pump Shaft								✔	✔		✔					
32-39768 Shaft								✔	✔	✔	✔					
32-01768 Shaft								✔		✔	✔					
34-92954 Pump Base												✔				✔
32-96768 Baseplate												✔			✔	
32-96742 Base												✔				
34-65786 Base												✔	✔			✔
34-65156 Base												✔	✔			
34-92941 Frt Cover													✔			✔
34-14236 Rear Cover													✔	✔		✔
34-14267 Cover													✔	✔	✔	✔
34-92461 Cover													✔	✔		✔
34-65427 Impellor														✔		
34-81454 Impellor														✔	✔	✔
34-85427 Impellor														✔	✔	
34-87654 Impellor														✔		✔

Figure 4.10 - Production Flow Analysis

At present, the team has not considered revising the processes. Some operations could move from NC to a manual process or vice versa. Revised processing could present interesting possibilities for tightening the relationship between product and process within each cell. It might relieve an overloaded cell. It might give additional work to an underutilized cell.

Gator Steel Foundry

Gator Steel is a jobbing foundry in northern Florida. Its specialty is small-to-medium-sized high-alloy steel castings. Gator also casts in mild steel. Many of Gator's castings become wear surfaces for earth-moving and logging equipment. The firm has about 800 active castings and uses a variety of molding processes. The work done by its managers and designers is an example of how a company that manufactures more than 100 items with complex processes accomplished Task 04.01.

Management decided to use C&C to rationalize and standardize the processes. They also wanted to set up work cells for their molding operations. After following the task procedure outlined in figure 4.5, they decided C&C was the most appropriate method for grouping products and processes into cells.

Because of the complexity of the casting process and the product range, Gator hired a consultant to help develop an appropriate coding system and carry out the analysis. He developed a polycode system, whereby each digit position represents the same type of information. This contrasts with a monocode system where the digit information varies based on the value of other digits. Such a polycode requires a longer code number and more computer memory and storage space. However, polycodes are easy to work with when conventional database managers are used.

Gator's consultant performed the work on a PC using a well-known database manager. The code carried information on both part features and current processes. The consultant used the part features to standardize processes and group parts into families.

Figure 4.11 illustrates a sampling of some of the 800 castings and their code numbers. The report header shows sequence number, part number, and name. It also shows the position number for each character in the code number. The code number has the following structure:

Position	Feature
01	Annual Usage
02	Lot Size
03	Finished Weight
04	Material

05	Heat Treatment
06	Smallest Window Size
07	Length
08	Tolerance Class
09	Binder and Sand System
10	Molding Equipment
11	Flask Size
12	Pieces Per Mold
13	Number Of Cores
14	Core-Making Equipment

CASTING GROUP TECHNOLOGY

SEQ #	PART#	NAME	POSITION 01	02	03	04	05	06	07	08	09	10	11	12	13	14	15	16
1	A3690	CLEVIS	3	2	1	1	0	0	0	B	0	0	0	2	0	0	0	H0
2	A59427	BRACKET	3	2	1	1	0	0	0	B	0	0	0	2	3	0	0	H0
3	A87655	CLEVIS	3	2	1	1	0	1	1	B	0	0	0	1	3	0	0	H1
4	A71123	STOP	4	2	1	1	0	1	1	B	0	0	0	1	3	0	0	H1
5	A83445	BEARING	3	2	2	1	0	1	1	B	0	0	0	1	3	0	0	H1
6	A19435	BEARING	2	2	1	1	0	1	1	B	0	0	0	1	1	0	0	H1
7	A14982	BUSHING	4	2	2	1	0	1	1	B	0	0	0	1	1	0	0	H1
8	A19400	PAWL	4	2	2	1	0	1	1	B	0	0	0	1	0	0	0	H1
9	A30045	BRACKET	3	2	1	1	0	1	1	B	0	0	0	1	0	0	0	H1
10	A52045	STOP	3	2	1	1	0	1	1	B	0	0	0	1	0	0	0	H1
11	A67112	END FITTING	4	2	1	1	0	1	1	B	0	0	0	1	1	0	0	H1
12	A97645	BRACKET	4	2	2	1	0	1	2	B	0	0	0	1	0	0	0	H2
13	A56719	LEVER	5	2	2	1	0	1	2	B	0	0	0	1	0	0	0	H2
14	A87235	LEVER	3	2	1	1	0	1	2	B	0	0	0	1	0	0	0	H2
15	A14561	BEARING	4	2	2	1	0	1	2	B	0	0	0	1	0	0	0	H2
16	A12334	RATCHET	4	2	3	1	0	1	2	B	0	0	0	1	1	0	0	H2
17	A34567	CONE	4	2	3	1	0	1	2	C	3	4	5	1	1	0	0	N2
18	A56112	CAP	4	2	3	1	0	1	2	C	3	4	5	1	2	0	0	N2

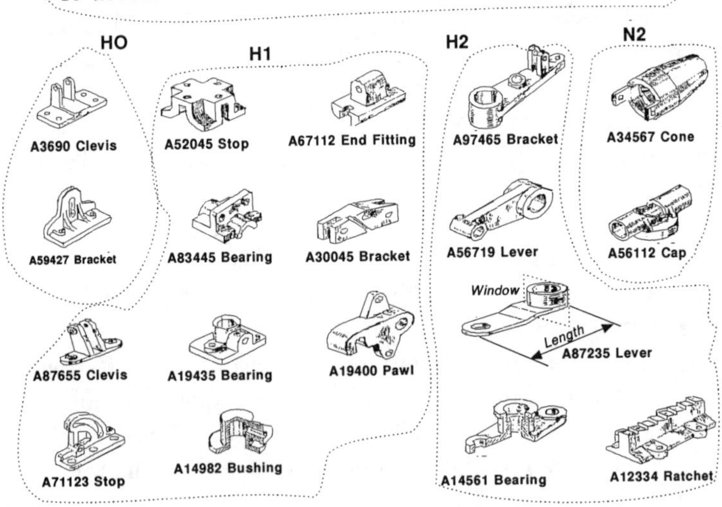

Figure 4.11 - Casting Coding and Classification

15	Secondary Core Equipment
16	Group Technology Family Assignment

Initially, the analyst coded each part using only positions 1 through 15. Position 01 is the first character of the code number. The number 3 in that position represents an annual usage of 201 up to 500 castings. *4* represents annual usage of 501 to 1,000 castings.

Position 05 is for heat treatment. *0* represents normalization at 1,600 to 1,675 degrees. This is one of several standard heat treatments at Gator. *3* represents a water quench followed by tempering at 800 degrees.

Position 10 shows the type of molding equipment. *0* is a Hunter Matchplate molding machine with a 24 by 24 matchplate. *4* represents an Osborne Jolt- Squeeze machine with a 16 by 12 flask.

Other character positions described size, weight, and other important characteristics. This information was vital for production planning, scheduling, costing, and tooling design.

The analyst then checked the database for obvious discrepancies. For example, certain molding equipment corresponds to certain size ranges. A retrieval on combinations of characters shows such inconsistencies.

Next, the analyst standardized the processes. Parts with similar features normally should have uniform processes. For example, high-volume parts should run on high-volume equipment. Any such discrepancies were corrected.

The analyst then began to group parts around similar processes, using character position 16 to identify each family. He assigned nine of the parts in figure 4.11 to the H1 family.

This family has parts with sufficient volume for a Hunter automatic molding machine. They use the same flask size and have the same material, allowing similar parts to run together and minimizing changeover time. Families H0 and H3 also run on the Hunter equipment but with a larger mold size.

The N2 family uses a no-bake binder system. The no-bake process is more expensive than green sand but achieves tighter tolerances. The tolerance class, Position 08, is different for castings in family N2.

Only a small sample of parts has been shown. The complete database, 816 part numbers, fell into 14 families. The number of parts in each family varied from 2 to 73.

In later steps, Gator Steel added process time and other information to their database. This allowed the staff to calculate average equipment utilization and people requirements for proposed work cells.

All code numbers in figure 4.11 are somewhat similar, and some are even identical. This is because the castings in the sample are similar in many respects. They all have about the same size and weight, and they use the same material. They have simple cores or no cores. Within each GT family, the code numbers have even more similarity.

Yet, from the design engineer's perspective, these castings are quite different: they have different names and functions and become parts in different products. From a marketing perspective, the castings also are different: they go to different industries and have different customers. From a financial perspective, the products have different pricing, overheads, and profit margins.

Designing the process

Task 04.02 designs or redesigns the cell's production process. This often results in significant improvements in productivity because it ensures that the process suits the work cell's part family. It also forces a critical review of the type, size, and capacity of equipment and may eliminate process elements that add no value.

This task is similar to some of the work in Task 03.11 at the macro-level but with more detail. At the macro-level, the process was defined only as far as was necessary to ensure a valid cell definition. All equipment or process times may not have been identified, and the designer may have recognized that further refinement was necessary.

Figure 4.12 shows the task procedure. In Block 1, a preliminary process and the equipment required for it is selected. This often is an existing process. For a completely new product, it may be necessary to start with a proposed process. This preliminary selection should result in a process chart and corresponding equipment list.

Block 2 refines the process. Non-value-added elements are eliminated and other options for equipment scale are explored. For example, a preliminary process for a steel fabrication cell needed a sophisticated cutoff saw—a large, high-volume piece of equipment that would serve five different cells. The process refinement examined the use of five smaller bandsaws. Each cell would have one of the saws. This eliminated five transport operations, reduced inventory and simplified scheduling. Process scale is a particularly important issue when making a transition from a functional to a cellular space plan.

Blocks 3 through 6 estimates set-up, equipment, process, and personnel times. Figure 4.13 illustrates the differences. Traditional time study often assumes that personnel, process, and machine times are the same. This assumption comes from an underlying supposition that one person operates a single machine, which often is invalid. Figure 4.13 shows that, after an initial set up, a person loads the machine and is then idle while the machine operates. That person then unloads the machine and inspects the part. The machine is idle during the inspection. The process time is the total throughput time for one part.

Disparities between person and machine times can generate important productivity increases. Where such disparity is significant, people can load a machine and then move to another process while the machine operates. The longer the process time, the more advantageous this becomes.

If set-ups vary significantly between parts, the setup time for each part and each operation should be estimated; otherwise, a blanket set-up time for each operation may be sufficient. Consider the possibilities for improving set-up times. These estimates do not have to be particularly accurate. Accuracy on the order of plus or minus 20 percent is sufficient. These times should be documented on a spreadsheet similar to Table 4.1.

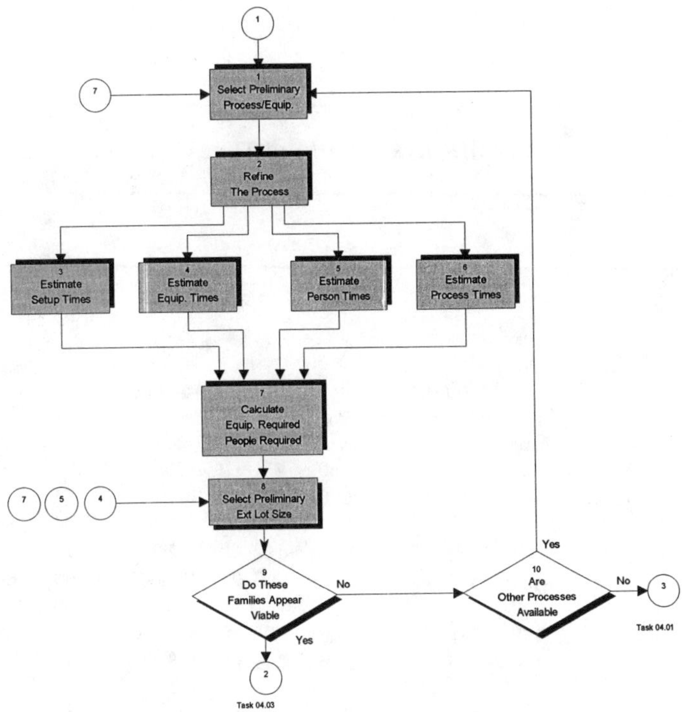

Figure 4.12 - Design and Refine Process

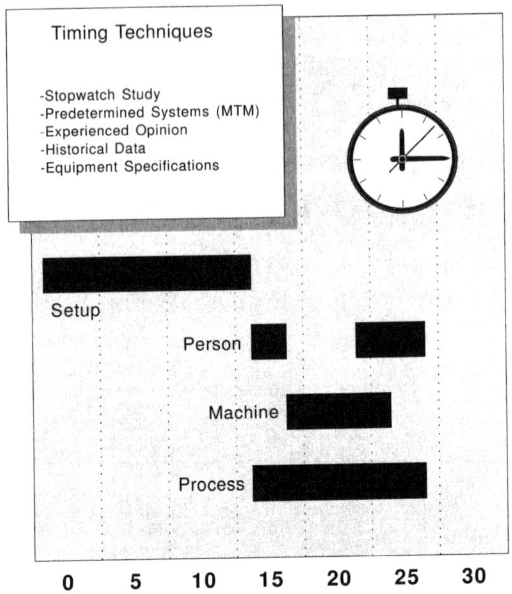

Figure 4.13 - Workcell Times

S/U per Day: 2
Units/Day: 88
Avail Min/Day: 420

	PER UNIT				PER DAY			
Station/Op	**Oper**	**Mach**	**Proc**	**S/U**	**Oper**	**Mach**	**Proc**	**S/U**
Head SubAssy	3.09	3.09	3.09	3.60	271.92	271.92	271.92	7.20
Diaphragm Assy	1.38	1.38	1.38	3.00	121.44	121.44	121.44	6.00
Body/Coil SubAs	0.88	0.88	0.88	2.00	77.44	77.44	77.44	4.00
Main Assy	2.19	2.19	2.19	1.75	192.72	192.71	192.72	3.50
Elect Assy	4.70	4.70	4.70	2.00	413.60	413.60	413.60	4.00
Elect Test	0.45	0.50	0.95	0.00	39.60	44.00	83.60	0.00
Flow/Press Test	1.03	15.00	17.06	2.00	90.64	1320.00	1501.28	4.00
Package	1.88	1.88	1.88	1.00	165.44	165.44	165.44	2.00
Totals	**15.60**	**29.62**	**32.13**	**15.35**	**1372.80**	**2606.56**	**2827.44**	**30.70**

People Reqd: 3.34167

TABLE 4.1

The time for each cycle when the equipment is busy and unavailable for other work is then estimated. These times may be obtained from equipment specifications, stopwatch timing, existing standards, or experience. The level of accuracy normally used to set work standards is not necessary. Accuracy of plus or minus 15 percent is usually sufficient.

How long is a person is occupied during an operation cycle? To calculate personnel time, the same sources used for equipment times are available, and the required level of accuracy is similar.

Process time is the time a part spends in process, not including queuing time. Some processes may have a curing or other wait, which is also part of the process time.

In situations where a process requires constant attendance, personnel, equipment, and process times may coincide. This also may happen when very short cycles preclude the possibility of a person moving to another operation.

Block 7 uses the time estimates to calculate the amount of equipment and the number of people needed for each process.

Block 8 selects the external lot size. External lot size is the quantity of product produced after a changeover or set up of the cell. This is usually different from internal lot size, which will be discussed later.

The form in figure 4.14 may be used as a guide for estimating the external lot size. Selecting an appropriate lot size requires intelligent balancing of three cost factors: set-up costs, inventory costs, and piece costs.

It costs time, effort, and money to set up or change equipment. This set-up cost must then be spread over the number of units produced. For example, it may take two hours to set up a machine at a cost of $10 per hour. The total set-up cost, then, is $20. If one production unit is processed, the setup cost per unit is $20. If two units are processed, a cost of $10 is added to each unit. If 100 units are processed, a cost of 20 cents is added to each unit.

If set-up costs were the only factor, infinitely large lots would be the result. Inventory costs are among the offsetting factors that limit the maximum lot size. Larger lots call for more inventory. Resources are expended to provide the space for that inventory, which also needs working capital and insurance against loss. It requires tracking and support systems to control it. Inventory also can become obsolete or damaged.

Piece costs are those costs that remain constant regardless of lot size. They include direct labor, materials, and some overhead. The more valuable the material and labor in a product, the faster storage and inventory costs rise.

When selecting external lot size, there are several principles to keep in mind:

1. Most firms use lot sizes that are far above the optimum. Designers should not assume that current lot sizes are realistic.

2. The total cost curve is usually flat near the optimum. Therefore, a significant variance from the theoretical optimum may have little cost effect. For this reason, choose a convenient round number or use a rule of thumb. Such a rule of thumb might be 0.5 days of production or a single customer order.

3. Conventional accounting systems often disguise many storage costs. These must be investigated with tenacity to make good decisions.

4. Set-up reduction can have dramatic effects on the shape of the total cost curve. It may become so flat that the lot size issue becomes irrelevant. Be sure to consider the value and effects of set-up reduction.

Diamond's A-team designs a process

The A Team at Diamond Equipment continued its work cell design project. Figure 4.14 illustrates the process chart for the original process and layout. Because the firm had organized itself along functional lines, this chart shows many moves, delays, and storages. The team proposed a new process based on the same equipment but in a cellular arrangement. Figure 4.15 shows this refined process.

Table 4.1 contains the time estimates. From these times, the team calculated the people and equipment requirements. They assumed four set ups per day knowing this might change when the external lot sizes were calculated.

The team then estimated the total set-up time within the cell at 15.35 minutes. The sum of all set-up times with the previous layouts was about 60 minutes. The team did not know exactly how to achieve this reduction, even though team members believed it was a reasonable goal. The cost for a single set up for the new cell was $4.32. This calculation is shown at the upper right in figure 4.14.

The team next calculated unit cost, which does not depend on set up or lot size. Direct labor, taken from table 4.1, was 0.25 hours. The labor rate with overhead was $18.67. Unit cost was, therefore, $56.71.

To determine storage cost, the team went to Diamond's accounting department, where the annual costs for operating the finished goods warehouse were identified. This annual warehouse cost, $236,901.00, is in figure 4.14. Team members determined the average number of diaphragm pumps in the warehouse during the previous year, and they determined the space necessary for each pump. They then assigned 0.003 percent of the warehouse cost to each diaphragm pump unit. The cost for storing a single unit for an entire year was $15.32.

Unit Cost

Item	Cost ($)
Labor (Hours * Rate) .25*18.67=	4.85
Material/Parts	49.86
Overhead	
Other:	
Total Cost/Unit	$ 54.71

Setup Cost

Item	Cost ($)
Labor (Hours * Rate) .25*16.87=	4.32
Setup Scrap	
Setup Materials	
Lost Production Profit (If @ 100%)	
Total Cost Per Setup	$ 4.32

Economic Lot Size

$$ELS = \sqrt{\frac{2.0 \text{*SU Cost(\$/SU)*Annual Usage(Un)}}{\text{Storage Cost (\$/Un)}}}$$

ELS = 104

Total Unit Cost

Lot	Piece	Setup	Storage	Total
25	54.710	0.173	0.010	54.893
50	54.710	0.086	0.020	54.816
75	54.710	0.058	0.030	54.797
100	54.710	0.043	0.040	54.793
125	54.710	0.035	0.049	54.794
150	54.710	0.029	0.059	54.798
175	54.710	0.025	0.069	54.804

Storage Cost

Warehouse	Cost ($)
Lease/Capital Charge	82,100.00
Labor (Hours * Rate)	112,020.00
Insurance	6,087.00
Maintenance	16,500.00
Taxes	3,400.00
Utilities	13,154.00
Obsolescence	1,540.00
OTHER	2,100.00
Subtotal	236,901.00
% of Warehouse/Unit	0.003%
Unit Cost * Internal ROR	8.21
Annual Warehouse Cost/Unit	15.32

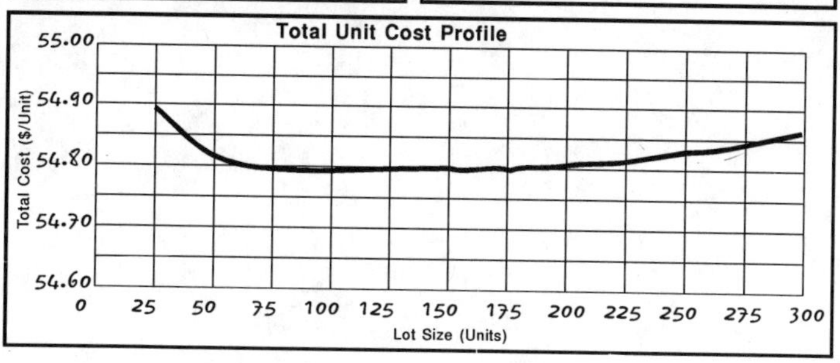

Figure 4.14 - Lot Size Analysis

day. This is equivalent to 19,360 units per year. Using this figure, the set-up cost, and the unit storage cost, they calculated the theoretic economic lot size (ELS). The ELS was 104 units, or greater than one day's production. The team knew that it had probably not identified all the storage cost and that total cost curves are often flat near the ELS.

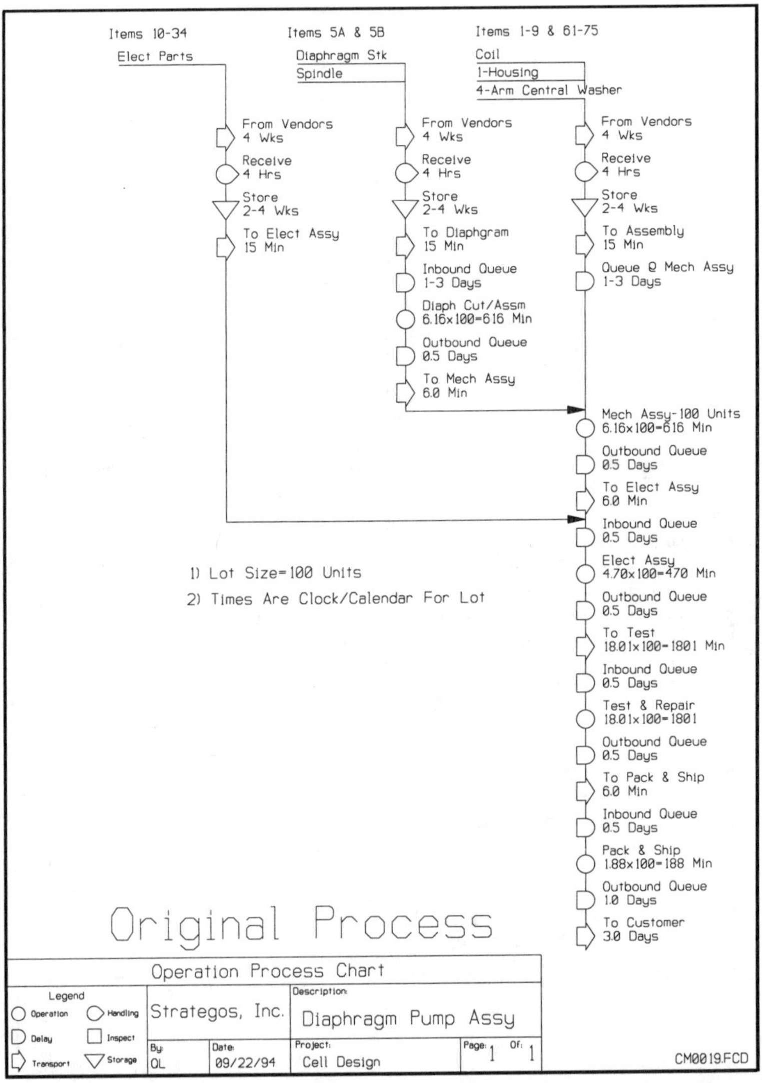

Figure 4.15 - Process Chart for the Original Process and Layout

This meant that a wide range of lot sizes brought almost the same costs.

They calculated total cost for a range of lot sizes that spanned the ELS. They plotted them at the bottom of figure 4.14. This confirmed that a lot size of half the ELS increased the unit price by only two cents.

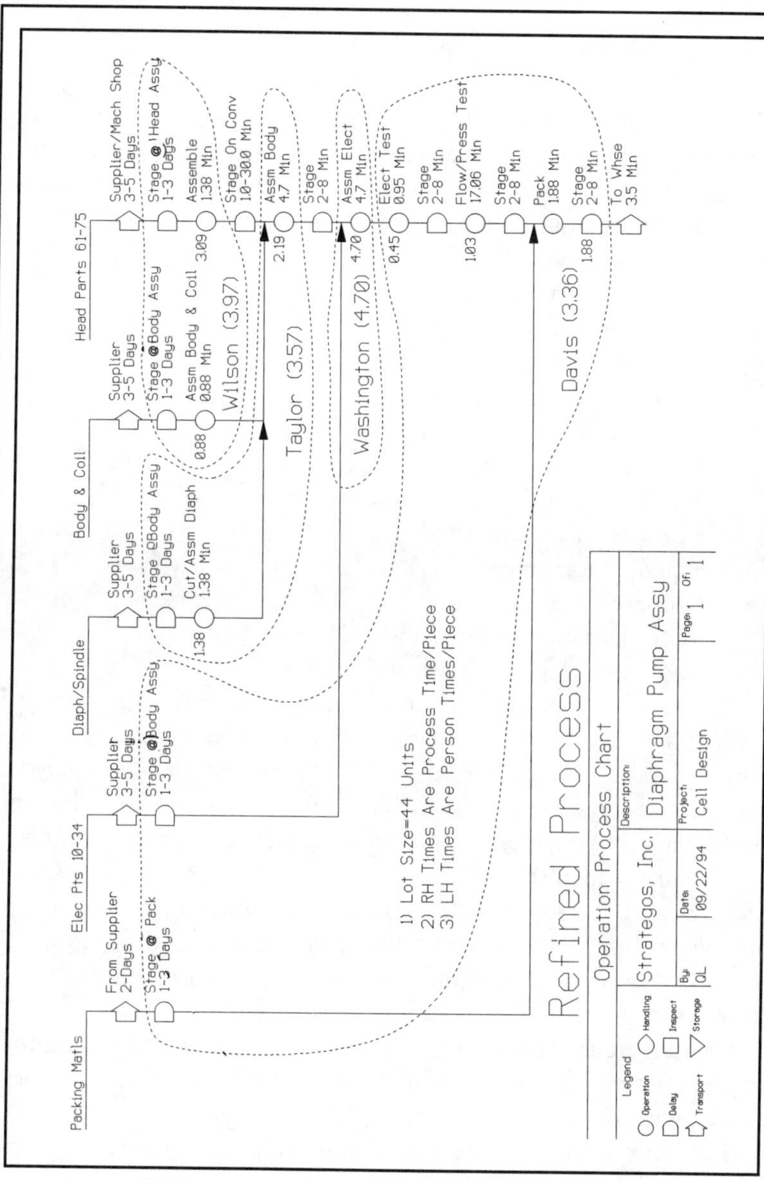

Figure 4.16 - Process Chart for the Refined Process and Layout

selected a lot size of forty-four units. This is one half-day of production, two set-ups per day. Based on the conventions outlined earlier in this chapter, this is a medium lot size.

They used two set ups per day for the spreadsheet of table 4.1. They also verified that the cell needed four people.

Planning cell infrastructure

Task 4.03 selects infrastructure elements; the procedure is illustrated in figure 4.17. Infrastructure elements support process without directly touching or contributing to product creation. Some infrastructure elements are tangible and some are not. The following infrastructure elements may be considered the minimum for adequate work cell design:

- external containers;
- internal containers;
- external material handling;
- internal material handling;
- external production control;
- internal production control;
- internal lot size (transfer batch);
- equipment balance method;
- people balance method;
- quality assurance;
- supervision;
- compensation system; and
- operator assignments and skills.

These infrastructure elements are an essential part of the work cell design. If a design team or designer does not select them, operators must do it ad hoc. Their selection seldom requires extensive analysis; however, it does require careful thought and experience. Each element must fit with the others into an integrated, well-functioning system.

The first step of the task procedure is the selection of the external containers and the identification of the size and type of container. Materials and products move to and from the cell in these external containers. Containers bringing material to the cell can differ from those removing products.

Containers can protect their contents. They can improve handling characteristics and convert small units to larger handling units. They can assist with visual control of materials.

The conventions described earlier in this chapter may be used to classify containers as single-piece, small, medium, or large.

Containers should never be larger than the lot size. Thus, large or medium containers should not be specified for a small lot size. This medium containers should not be specified for a small lot size. This same sort of incompatibility can occur with other design elements.

After the container selection, the structure of its route and the type of handling equipment that will be used are specified. This is Block 2 of the task. Handling often changes significantly when manufacturers move from functional to cellular layouts. The new system must accommodate smaller loads and more frequent moves. This affects the container route structure and the type of equipment employed.

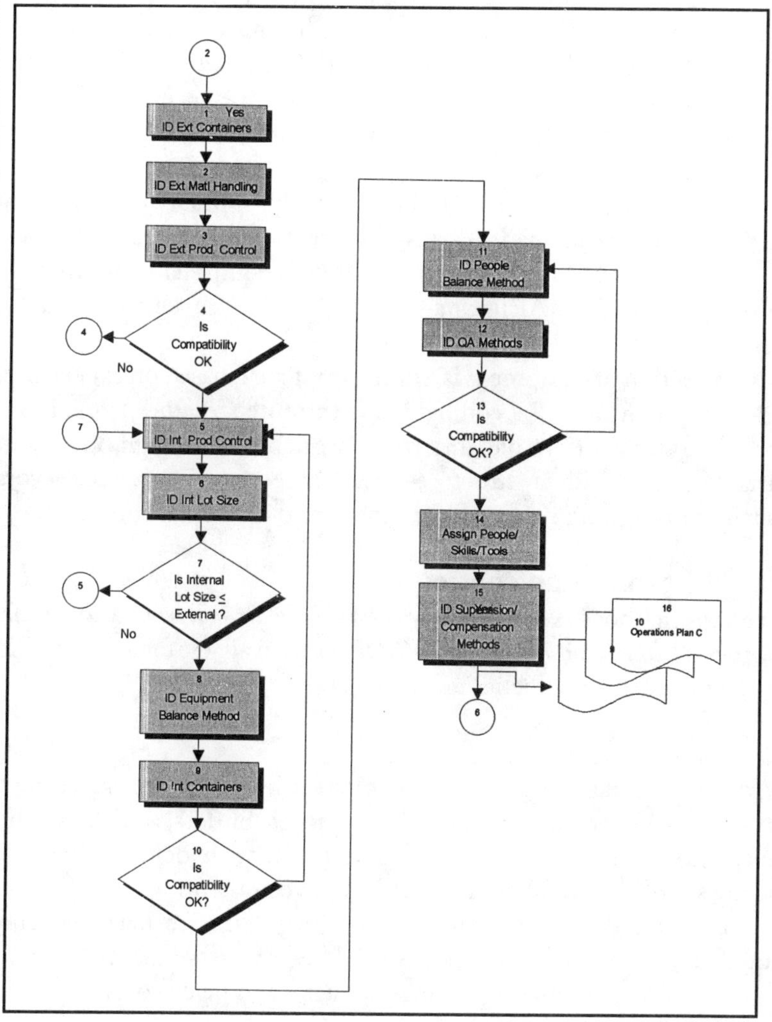

Figure 4.17 - Task 04.03 - Identify Infrastructure Elements

Figure 4.18 shows the basic route structures available. (A) is a direct route structure, where material moves from its origin directly to its destination. Such a structure fits well with large lots and functional layouts. Full loads for the equipment employed are necessary. Fork trucks moving pallets usually operate on a direct route structure.

In (B), a channel route structure is shown. Channel structures fit with smaller containers and less-than-full loads. The equipment moves along a fixed route. At each stop it picks up loads originating at that point and drops off those destined for that point. City buses use such a structure.

For situations where control of movement and material is critical, a central system (C) works well. All loads move first to a central location. A second move takes them to their destination. Central systems work well where sorting or storage takes place. Federal Express used a nationwide central system for many years. Many warehouses and distributors use central systems.

A s is illustrated in (D). This structure works well with long distances and partial loads. Loads move first to a terminal. They then move to a second terminal, often on different equipment, and then on to their destination. Airlines and long-distance truckers often use a terminal system.

Selection of equipment is, for the most part, based on experience. The equipment must be compatible with route structure, containers, and flow rate. For example, fork trucks work best with high flow rates, variable routes, direct structures, and large containers. Conveyors usually suit high flow rates, fixed routes, and direct structures.

External production control

Five basic methods—*direct link*, *broadcast*, *kanban*, *material requirements planning (MRP)*, and *reorder point (ROP)*—are available for coordinating production. This is Block 3 in figure 4.17.

Direct link

With direct linkage, processes have a physical link. Product moves from the first to the second process without queues, buffers, or delays. The processes start and stop together. They operate with identical lot sizes, usually single-piece. There is no WIP between them.

Direct link is only feasible when processes are balanced and require about the same time per cycle. The processes must be capable of synchronization, have the same products, and be co-located.

Broadcast

A broadcast system is slightly more flexible than direct link. In this system, a schedule dictates the rate and product mix for the final operation. The same schedule is applied to upstream operations, which make their components in the same order (the line set order) as the final operation. They deliver these components just ahead of the scheduled assembly time. Co-location is not necessary; however, the capability to make the same lot size—usually a single piece—is. Broadcast systems often are seen where subassembly operations feed assembly lines.

Kanban

Kanban systems use a small mixed stock. The stock point is often at the producing work center. This stock has enough of each item for a few hours to a few days of production. Users of these items pull small quantities from the stock frequently. They signal the producing work center at the time they make each withdrawal. Sometimes this signal device is a card, hence the use of the word kanban, which is Japanese for card.

At the producing work center, people track the incoming cards and observe stock levels. They then schedule their production to minimize changeover efforts while replenishing the stocks. Kanban systems sense and respond to changing demand very quickly. They operate with very low inventories and need short set-ups and small lot sizes. It is unnecessary to coordinate production exactly. A work center may feed and draw from many other work centers. The system is simple and does not need computers or sophisticated communications equipment. Kanban is often a good choice for coordinating work cells.

Material requirements planning (MRP)

MRP systems use computers to schedule and track production. They also can plan capacity, collect costs, and provide many other functions. A basic MRP system has databases with the following information: bills of materials, routings, inventory, and lead times for each operation.

The system works from a forecast of end-product demand and the above information. It produces a schedule for each work center. Feedback updates the system as execution takes place. MRP systems are very flexible and can schedule virtually any type of shop or operation. In theory, a modern MRP system is almost the ultimate in scheduling.

Practice is somewhat different. To work correctly, bills of material must be almost perfect. Routings must be current. Inventory accuracy must be 99 percent or better. Lead times must be realistic. Variation from lead times must be minimal, and the forecast must be reasonably accurate. Such conditions rarely exist.

Reorder point (ROP)

ROP systems maintain a significant stock of each item. As users withdraw items, the remaining stock declines. When the stock reaches a pre-determined level—the reorder point—a signal initiates a purchase order or shop order to replace the stock. Ideally, the remaining stock lasts until the replacement quantity arrives. ROP systems are simple. They can operate manually, semi-manually, or with a computer. When demand is steady and replenishment times do not vary, ROP systems work well. These conditions rarely exist.

ROP systems are superficially similar to kanban. They differ in the following ways:

1. ROP systems generate a replenishment signal only at the reorder point. Kanban systems generate a replenishment signal with each withdrawal.

2. Compared to ROP systems, kanban systems operate with frequent replenishments and small lot sizes.

3. ROP systems usually take days or weeks for a replenishment signal. The feedback for a kanban withdrawal is very fast, often minutes or hours.

4. ROP systems usually have a separate warehouse or scheduling system. Producing workers usually operate a kanban system.

5. ROP systems work well for small, low-cost items such as hardware, where simplicity of operation overcomes the inherent inefficiency of the system.

There is a hierarchy of methods:

1. direct link;

2. broadcast;

3. kanban;

4. material requirements planning; and

5. reorder point.

When conditions allow, direct link is the most desirable method. ROP is the least desirable. As work cells are being planned, it is best to use a system as high in the hierarchy as possible. Use ROP only under very special conditions. Work cells generally work best with direct link, broadcast, and kanban systems.

Within a facility or work cell, several of these systems may exist at the same time. Kanban may work for certain processes and products, and MRP for others. Kanban might be used internally and MRP for purchased items.

Over time, the system may change. A process-focused plant might initially have an MRP system. Management may begin to connect operations with kanban systems for selected products. With a new space plan, operations would consolidate in work cells or have direct linkage.

Block 5 of Task 04.03 identifies internal production control, which addresses the issue of coordinating processes within the work cell. It helps operators decide what to make and what operation to perform next. The selection of an internal production control method usually ties to the selection of balance methods. Within the cell, several production control methods may be used for different parts of the cell.

Part of the discussion of external production control applies internally. Direct link, broadcast, and kanban often work well within cells. MRP and ROP usually work only in functional or process-focused work cells.

Circulation is also a viable option. This method uses mobile operators to carry a product or transfer a batch through each operation. If the cell requires several operators, they follow each other around the cell. Circulation also balances people. Because each operator does every job, he or she always has the same workload.

Here are some guidelines for applying the methods to internal production control:

- use direct link for balanced and synchronized operations.
- use kanban where the cell has short process times, heavy equipment use, or long distances between operations.
- use circulation for portable products where distances are short. There also should be inherent equipment balance or excess capacity on some equipment.

After the internal production control is identified, the internal lot size is identified. This is Block 6 in figure 4.17. The internal lot size, also known as the transfer batch size, is the number of units processed at each operation before transfer to the next operation.

The internal lot size usually is smaller than the external lot size in a work cell. This brings many advantages: work flows more quickly, the amount of WIP is smaller, quality feedback is rapid, and response time is fast.

The smallest internal lot size consistent with certain limiting factors is desirable. The ideal internal lot size is one-piece.

Equipment batching is one factor that might limit selection of a small or one-piece internal lot size. Some equipment may need a significant batch for proper or efficient operation. For example, a rotary shot-blast machine may need many parts to avoid excessive wear and to offset a long cycle time. A machining fixture may be designed to hold several work pieces.

Cycle time and transfer time also may be limiting factors. Some equipment, for example, may have a very fast cycle of, say, 0.1 minutes. The time to move a single work piece to and from the equipment may approach this cycle time. In such situations, operators spend so much time transferring one-piece lots that their efficiency drops and equipment utilization is unacceptably low. In this case, transferring in batches of twenty pieces would give a cycle time per batch of two minutes.

The workload on equipment in a cell must have some degree of balance. This improves cell productivity and equipment utilization and smoothes work flow. Block 8 identifies the equipment balance method.

Balance can be static or dynamic. Static balance means that equipment times for each operation, averaged over the entire product mix, are similar or the same. Static balance prevents one or more operations from becoming a bottleneck when other operations are not being utilized. A high degree of static balance is necessary when the cell contains more than one piece of high-cost or critical equipment. For workbenches, typewriters, or other low-cost equipment, it matters little if there is low utilization.

Dynamic balance is the balance between operations from work piece to work piece. The process time for operations may vary. Differences in product may cause this, or the process may vary inherently.

Dynamic imbalance can reduce throughput of the entire cell because of interference between operations. For example, operation 1 has completed a work piece but operation 2 cannot accept it immediately. At other times, operation 2 may be ready for the next work piece but operation 1 is not finished. This reduces capacity of the cell and can occur even when the average balance appears good.

There are three methods for correcting static or dynamic imbalance: queuing, inherent, and excess equipment. *Queuing* uses short queues between operations. This ensures that downstream operations always have a work supply and that upstream operations always have a destination for their work. Queuing can compensate for dynamic imbalance, but it cannot compensate for long-term static imbalance.

The *inherent method* employs processes with low process time variability and identical process times. Thus, the balance is inherent in process selection and does not need a compensating mechanism. Inherent balance usually is difficult to achieve.

Excess equipment uses surplus capacity to compensate for imbalance. In this method, the work cell is designed around high-cost or critical equipment. Low cost or secondary equipment with surplus capacity is sized and selected. The consequence of imbalance, idle equipment, then can do little harm. This method is also useful where utilization of people is more important than equipment utilization.

Block 9 identifies internal containers, based on the conventions mentioned previously. While internal containers have a somewhat different role than the external containers, their purpose is the same. If the internal lot size is greater than single-piece, the container should match it. This facilitates transfer and makes it easy to track internal lots.

A work cell should have some means to ensure that each person in it has about the same work time and effort level. This is the issue of balancing people in the cell (Block 11). The available methods are: queuing, circulation, fixed assignment, and float.

Queuing not only balances equipment, it also balances people. The method works well with short process times and varied product mix and may be used with fixed or floating task assignments.

Circulation applies when there are either very short or very long internal idle times and is based on the difference between the person time and the equipment time. With a very short internal idle time, the person remains with his work piece. With a very long internal idle time, fifteen seconds or more, the person initiates the process for a particular piece, picks up the previous piece, and then moves to the next process. It is a "leapfrog" procedure. The next person to arrive substitutes a new work piece for that just completed and moves on with it.

Circulation also works well when there is excess capacity on secondary equipment. Here, the secondary equipment may remain idle for much of the time.

To be effective, circulation must have short travel distances and people who are competent in all the processes.

As a balancing tool, fixed assignment requires inherently balanced processes. One operator has a fixed workstation. Fixed assignment often is used with queuing for equipment balance and production control.

The floating system for people balance assigns an operator a primary and secondary task. The secondary task may be a primary task for a neighboring operator. Each operator performs the primary task until signaled to move to the secondary. This signal may be as simple as a verbal call for assistance from a neighboring operator who is falling behind. Queues can be an effective signal. A queue that becomes too short or too long is the signal for one or more people to move to their secondary tasks. They then work down the unnecessarily long queues. This method calls for skilled people, teamwork, and team spirit. It can be highly satisfying for the team members.

Deciding which quality assurance method(s) to employ is done in Block 12 of figure 4.17. Quality assurance should be an integral part of the process. Three approaches are available: inherent, inspect/reject, and SPC/TQM.

The *inherent approach* ensures quality by selecting processes that inherently produce few or no defects. There is little point in elaborate quality inspection when defects are negligible and inconsequential. This situation is rare, but it does occur. The inspect/reject method introduces a formal inspection for quality. When defects occur, the product is repaired or scrapped. This is often used with a separate inspection department and a policing approach.

The preferred quality assurance methods use *total quality management* (TQM) often combined with *statistical process control* (SPC). In this method, the workers monitor their own processes and quality. The emphasis is on prevention and process control. TQM/SPC works especially well in a cellular environment.

Inspection points, procedures, and quality standards should be specified. This is an integral part of cell design and should not be assigned to an outside quality department.

As part of the cell design, operators should be assigned to tasks (Block 14). Task assignments may be fixed, floating, or circulatory. A fixed task assignment allocates an operator to a particular workstation or set of tasks. This is the traditional "one person-one machine" approach. If the entire cell has fixed assignments, there must be inherent balance.

Floating task assignments give an operator a primary set of tasks and a secondary set. As the workload fluctuates, the operator may move from the primary to the secondary. For example, operator 1 may have a set of tasks that, on average, occupy 80 percent of his or her time. operator 2, immediately adjacent, has a primary set of tasks that overload him or her by 10 percent. In that situation, operator 1 might be given a secondary assignment to assist operator 2 from time to time.

Operator mobility and a task design that allows for assistance are integral to a floating assignment. A mechanism for signaling the assisting operator when it is time to change tasks is also necessary. The signal can be a verbal request, a marker on a queue, lights, or some other method.

The circulation method balances equipment and people within the cell, but it also tells the operator which task to perform.

A particular work cell may use one or all of the above methods of assigning tasks to operators. The assignment begins with the refined process chart and the element times spreadsheet. For each operation, it must be determined whether a person's time or the process's time will govern. For example, a large differential that allows a person to move to another operation should use the former. These times are noted on the process chart.

Next, planners divide the total times for all elements by the number of workers. This is the number of minutes of work for each person if the work is perfectly balanced. It should be used as a target for assigning tasks. Then, the maximum work for any one person is determined. This is the available time in one day divided by the design volume; it is also the average cycle time for the cell.

The tasks are then assigned. Those that occupy more than one person's time are divided, and those that occupy significantly less than one person's time are combined. When dividing and combining, designers look first at adjacent tasks and then for those that might occur on opposite sides of a U-shaped cell. When several tasks are assigned to the same person, that person must have the ability to move between them on the layout.

Each person must have compatible tasks. One way is to concentrate special skills such as welding capability. While full cross-training is the ideal, there will be occasions when the cell will have to work without a full set of cross-trained people.

The final step in Task 04.03 is identifying the methods of supervision and compensation. Supervision in a work cell can enhance or disrupt the cell's performance. This issue is especially important when making the transition from conventional to cellular manufacturing. Two methods of supervision are: command and control and self-managed teams.

Command and control is the traditional approach developed from the Roman Legions, the Prussian army, and the Industrial Revolution. Supervisors think and give orders; workers follow the orders without question. The system works well for fielding a massive army of rabble. In industry, it can be effective for simple tasks that use uneducated and unskilled workers.

The most successful organizations in many fields of endeavor, including the United States army, no longer use unquestioned obedience as a guiding principle. Such organizations nearly always use some level of empowerment and participation.

A methodology that uses empowerment and participation is the *self-managed work team*, also referred to as a cybernetic work team. Such teams organize and execute their own activities. This, however, requires experience, training, and appropriate management.

Many work cells require a certain level of teamwork for maximum effectiveness; therefore, properly designed work cells fit nicely with cybernetic work teams. They can enhance teamwork and the feeling of team responsibility.

Cells that require teamwork for proper functioning are ineffective if team members receive no training, and supervisors do not allow them to make the mistakes that team development requires. Management sometimes assigns supervisors with no training and little temperament for teamwork. Even with proper training and coaching it takes a team three to eighteen months to mature fully.

In some organizations, management is unwilling to dedicate the resources and patience required for team-based cells. In this situation, the cell designers should recognize this reality and design their work cells to function with conventional approaches to supervision.

Compensation affects quality, productivity, and teamwork. While compensation systems vary, most firms use one or a combination of the following types: hourly or salary wages, individual incentives, group incentives, pay-for-knowledge, and hybrid systems.

Hourly or salary wages is the traditional compensation system. People receive pay for the hours they spend at work, or they receive a fixed pay each period with an expectation that they will spend a certain minimum number of hours at work. This works reasonably well in almost all situations but depends on other factors for motivation. It is team spirit, self actualization, or fear of punishment that motivates people to be productive, learn more about their jobs, and carry out their work with quality.

Combined hourly wage and salary systems often pay workers by the hour and managers with a salary. This approach sets up a two-tier class system that may cause friction and rivalry within the organization. Nevertheless, people are familiar with these systems, and they often are a reasonable choice.

All-salary systems are a good choice for team environments. Here, everyone is on salary. This helps reduce the divisive class structure of managers versus workers. Since the salary is largely independent of incremental cell output, productivity and quality incentives must come from teamwork and self-actualization.

Individual incentives base all or part of a person's pay directly on his or her task output. Such systems reward output quantity, but they often inhibit teamwork and usually have a negative impact on quality. They reward job knowledge only for the task at hand. Individual incentive systems work well only where narrow jobs require a single worker. They require a process with high intrinsic quality or a product that is easily inspected for quality.

Some of the difficulties with individual incentives disappear with a group incentive system, where members of small teams receive the same pay based on the team's output. This encourages teamwork and allows people to expand their job knowledge within the team. However, group incentives rarely help promote quality. In fact, they sometimes encourage conspiratorial behavior.

Where the work environment requires significant skills, a *pay-for-knowledge system* may be appropriate. In this system, people receive hourly wages or a salary based on the range and depth of their skills and knowledge. This system fits well in many cellular environments where team efforts and cross-functional skills are required.

Pay-for-knowledge systems are not easy to implement. Such systems can cause teamwork dysfunction as members vie for job experience that will enhance their pay. Productivity and quality may become secondary considerations.

Some firms may attempt to use hybrid systems of compensation with a baseline pay rate dependent on seniority or job position. Superimposed on the base pay are incentives for productivity, quality, and skills. Such systems are complex, often generating unintended consequences. Individuals may attempt to manipulate the system to maximize their pay.

Space plan designers should understand the effects and consequences of compensation on the operation of their facility. For example, a team-based cellular structure is unlikely to work well in an individual incentive environment. Pay-for-knowledge systems have little purpose in a functional facility.

The work cell operations plan

As infrastructure elements are chosen, designers must begin to visualize a layout. All the infrastructure elements connect with each other as well as product selection, process, and space plan.

Figure 4.18 shows a work cell operations plan. This form is a convenient place to document infrastructure decisions. It is also a reminder of the full range of choices for each element. The operator assignment and skill matrix area shows job assignments and skill requirements for each operator. Operators are listed along the left side, skill categories and operations across the top. For each operator, a solid bullet represents each primary operation and primary skill requirement; an unfilled bullet represents secondary skills and operations.

The A-team selects infrastructure elements

The A Team has completed product selection and defined its process

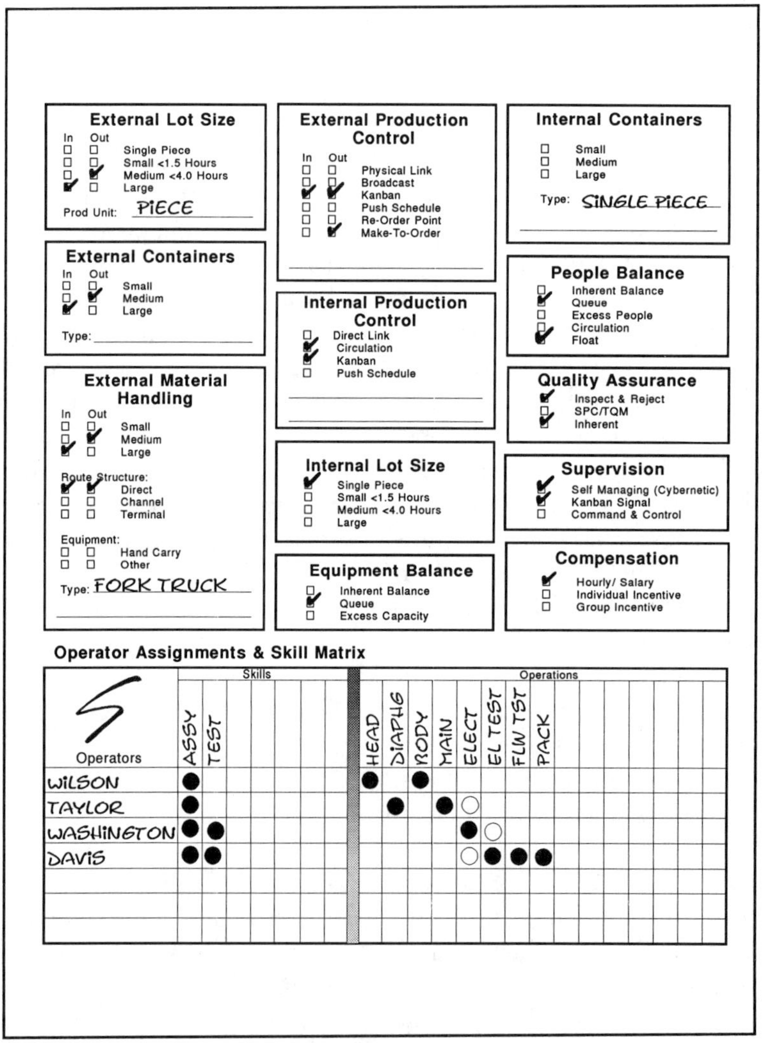

Figure 4.18 - Workcell Operations Plan

for the assembly of diaphragm pumps. Figure 4.18 is its work cell operations plan. The team has selected a medium external lot size of forty-four pieces. This is the lot size for finished pumps, but the team members realized that purchased items and items from other departments may come to the cell in larger lots. Figure 4.18 reflects this.

The team decided to package the pumps in cartons, which will be placed on pallets, a single lot for each pallet. This is the outbound container. It is a medium container, consistent with the medium lot size. Inbound material also comes on pallets, most of which will hold parts for more than four hours of production. The inbound pallets are, therefore, large.

For external material handling, the cell needs medium and large equipment corresponding to the container size. The team decided to use fork trucks for both inbound and outbound handling. Because fork trucks carry a full load (one pallet), the team selected a direct route structure. Other possible route structures are shown in Figure 4.19.

The kanban method was selected for both inbound and outbound external production control. Occasionally, the cell will make special orders. The team decided to use the kanban and direct link methods for internal production control.

The cell will balance equipment with queuing and excess capacity. The team decided not to use containers internally since the assemblies and components are easy to handle, and the internal lot size is single piece.

People in the cell will balance their work load using queuing and float systems. Quality assurance is inherent in some operations, although the electrical and flow tests are inspect/reject.

The four people in the cell will operate as a cybernetic work team and use kanban signals for cell management. Compensation is hourly/salary.

Designing a space plan

Task 04.04, "Design Cell Layouts," (fig. 4.20) arranges equipment, people, and other physical elements of the work cell. As noted in Chapter 2, every space plan has four fundamental elements (SPUs, affinities, space, and constraints). At the micro-level, using the fundamentals is usually simple and straightforward (see fig. 4.21).

SPUs in the micro-space-plan are usually workstations, equipment, or other significant features. These are derived from the equipment list or process chart. Diamond's A Team identified the ten SPUs shown in the affinity chart of figure 4.21.

Affinities are straightforward at the micro-level. Examining the process charts and work cell operations plan usually will identify them. Using simple judgments, space planners note the affinities and their underlying reasons on the affinity chart. Typically, there will be many A and E affinities and few I and O affinities. With experience, on simple micro-space-plans, a designer may move directly from the process chart to a layout.

Space, the third fundamental element, is also important. The designer usually moves directly from a configuration diagram to a scaled

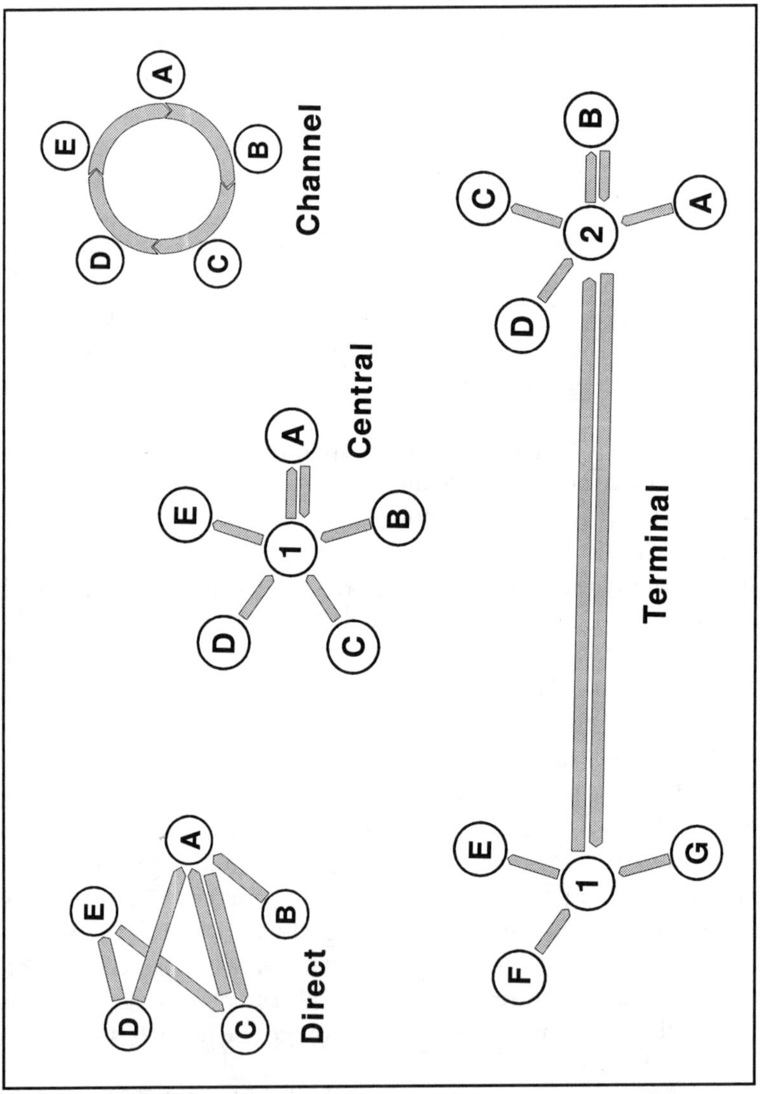

Figure 4.19 - Material Handling Route Structure

space plan. Occasionally, equipment templates may be superimposed on the configuration diagram. Highly flexible SPUs, such as staging or queuing areas, can be sized as the design proceeds.

Because the focus of micro-space-planning is limited and specific, a formal constraint summary seldom is necessary. Knowledge of equipment and process is sufficient to incorporate constraints as the designers proceed with the space plan.

Work cell shapes

As the layout progresses, it will probably take on one of several classic shapes. Among the most common shapes are *straight line*, *serpentine*, *U*, and *inverted-U*. Henry Ford's original assembly line was straight line

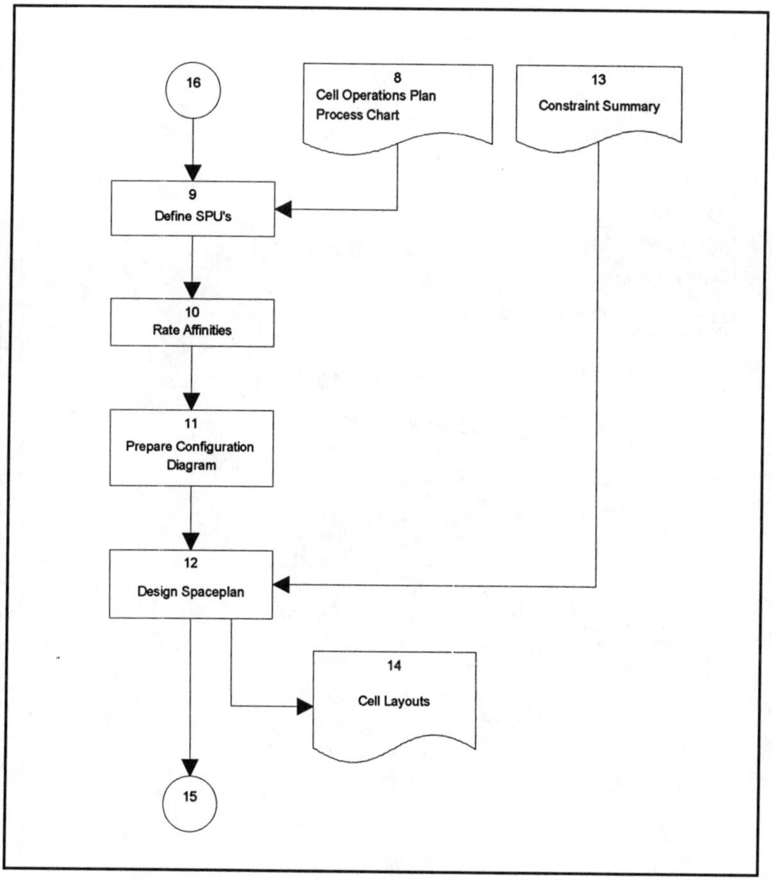

Figure 4.20 - Task 04.04 - Design Cell Layouts

(fig. 4.22). The work product moves from one end to the other in a straight line. It passes or stops at each workstation. When Henry Ford and Charles Sorensen first devised this arrangement, they were solving a material flow problem. Many parts went into a relatively small product where a large number of people worked at short-cycle tasks. A straight-line cell is still well suited to such situations.

Straight-line arrangements do not allow easy communication between people. Balance is difficult because the opportunity to float is limited. Once balance is achieved, it is difficult to maintain if the line speed or product mix changes. Long, thin line arrangements are often

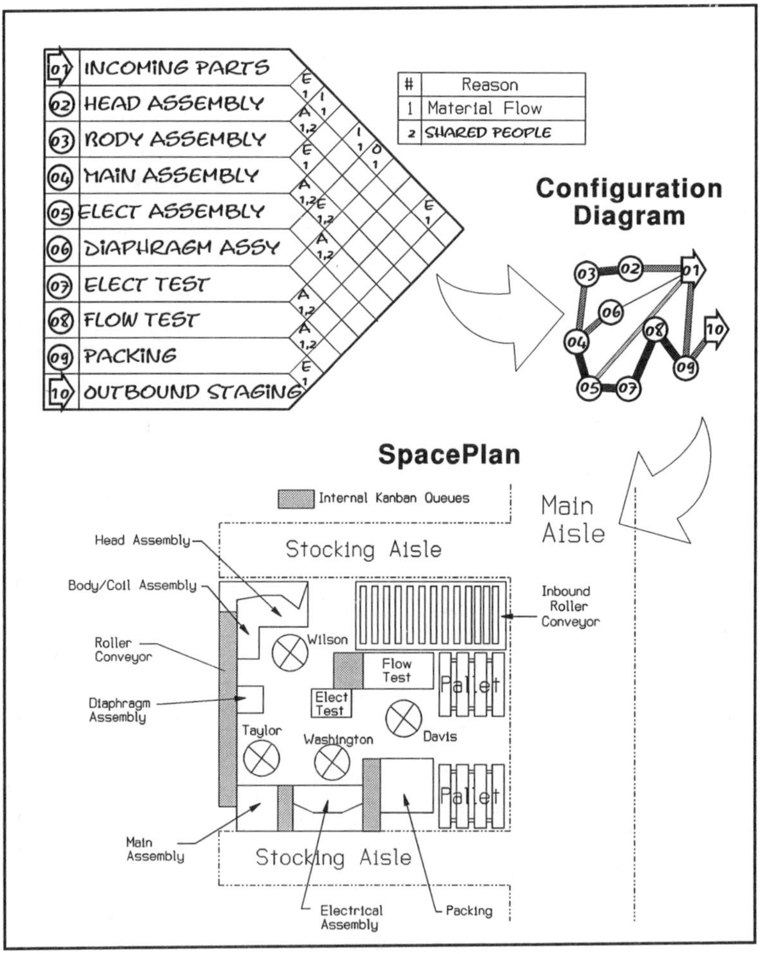

Figure 4.21 - Diaphragm Workcell SpacePlan

difficult to fit into existing buildings and macro-layouts.

Serpentine cells (fig. 4.23) take long production lines and weave them back and forth to fit building and layout features. Such an arrangement is more compact than the straight line cell. It allows better communication and somewhat better zone floating.

Serpentine cells sacrifice the ideal material flows of the straight line arrangement. They offer less storage at the line.

U-shaped cells (fig. 4.24) are highly flexible because they allow people to rearrange their work patterns easily. They work with floating assignments, circulation, and fixed assignments. As volume or product mix changes, it is easy to alter the number of people or their work assignments. It also is easy for people in the cell to communicate.

Material flow is sometimes more difficult in a U-shaped cell because material must be brought to the interior of the U. This arrangement may not work well for intensive assembly work with many different parts and significant on-line storage.

U-shape cells are limited in size. As the U becomes larger, communication between workers diminishes. The ability for workers to assist each other and interchange jobs also diminishes.

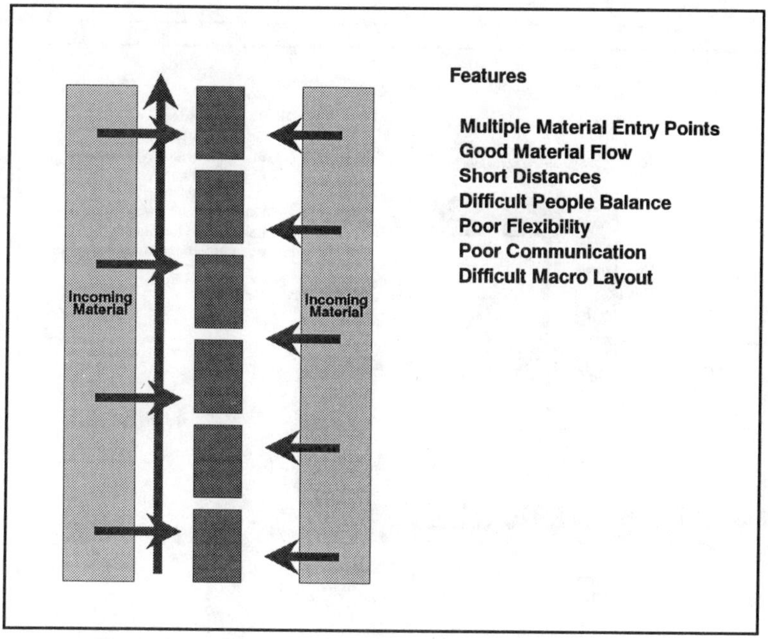

Figure 4.22 - Straight Line Cells

The inverse U (fig. 4.25) is a variation that allows more line storage and reasonably good communication but is less flexible than the U-shaped cell.

Actual cells can have many variations and combinations of these basic shapes. When the amount of work required on a product dictates many workstations, it may be best to link a series of cells, where each cell feeds the next. Within each cell, many of the advantages of the smaller, self-contained operation still exist.

Micro-space-planning for teams

Teams, in their various forms, are one of the most important developments in the American work place since the Industrial Revolution. Their freewheeling structure suits current needs for wide information exchange. They pool the knowledge of all members.

Teams can replace monetary motivation with intrinsic motivation. They can fulfill the higher needs of Maslow's Hierarchy: social, self-esteem, and self-actualization.

Teams come in a great variety. They may be temporary or permanent and may include members from all levels of the organization. Cross-functional teams bring people together from different parts of

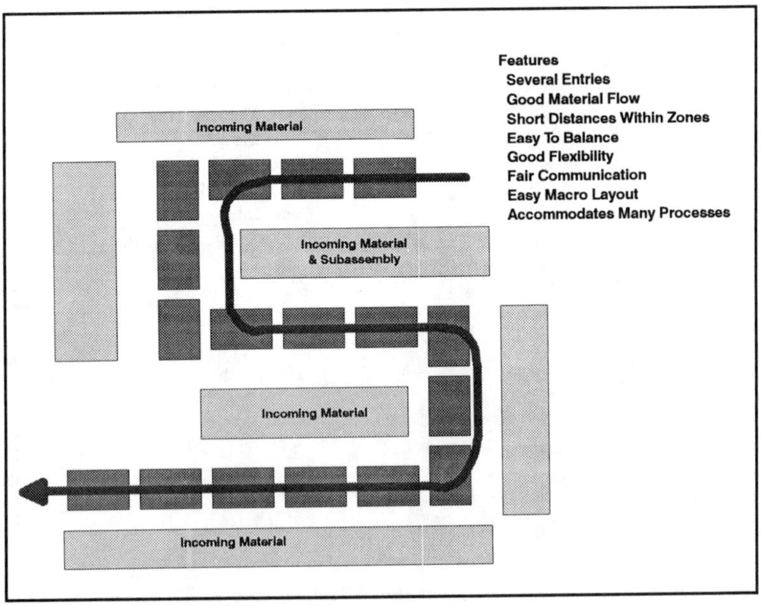

Figure 4.23 - Serpentine Cells

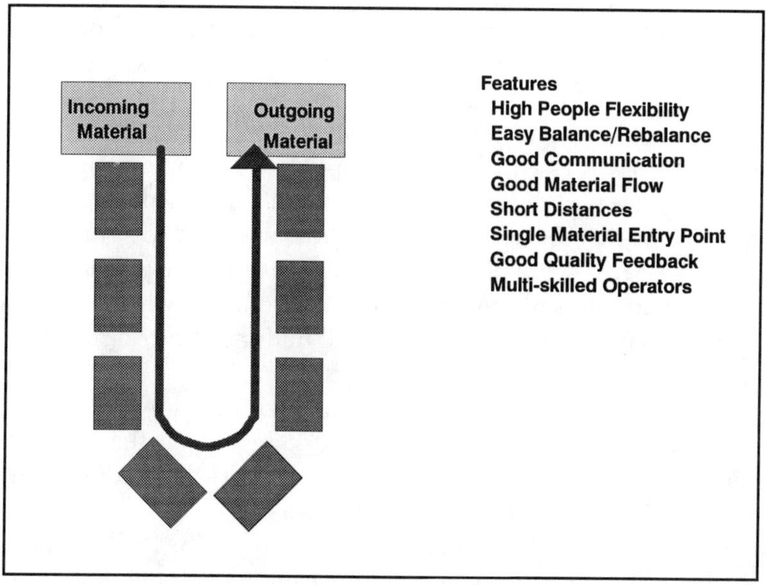

Figure 4.24 - U-Shaped Cells

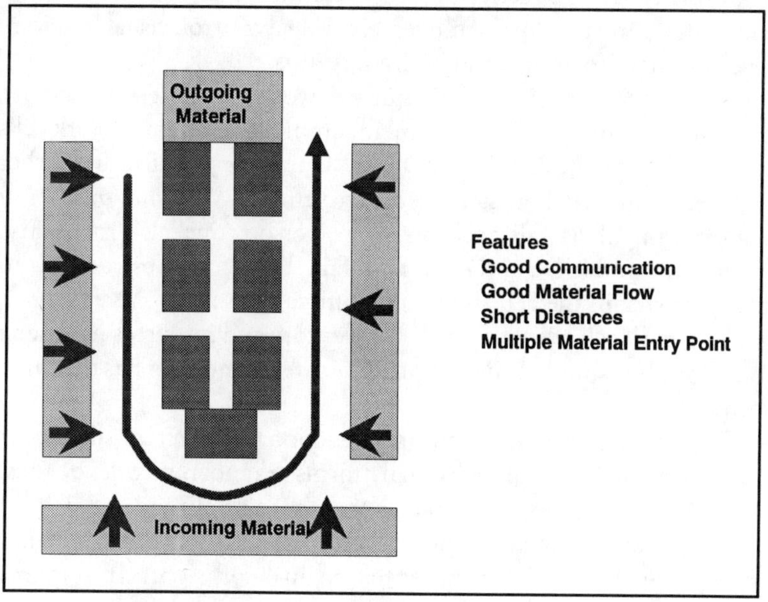

Figure 4.25 - Inverse U-Shaped Cells

functionally oriented organizations. Problem-solving teams assemble to delve into a particular problem or series of problems. Work teams function together in their daily tasks.

Work teams are especially important from the facility planning perspective. A space plan can promote or inhibit teamwork. Conversely, close teamwork is essential for the proper functioning of many space plans. If a space plan dependent on teamwork is designed, the designer should ensure that the teamwork capabilities will be in place to support it. Otherwise, the space plan will not perform effectively. Conversely, a firm that attempts to use teams in an environment where tasks and people are isolated and independent will find the efforts difficult and disappointing.

Cellular operations fit especially well with work teams where a small group, usually two to ten persons, works together in the same area. Often these teams are self-managed. With minimal direction from a supervisor, they plan and conduct their own work.

The members of a self-managed work team (SMWT) must be trained to acquire necessary interpersonal and managerial skills. Such teams go through a dynamic, but predictable, growth cycle before they are effective. During this maturation, they need skilled coaching and patience from management.

Work cells that effectively promote teamwork have certain features. Among these are: freedom of movement, visual control, visual freedom, conversational proximity, and bulletin boards.

To assist each other with normal work and interruptions, to communicate, and to cross-train, members of team-oriented work cells must be able to move between tasks. Conveyors, walls, and other equipment can make such movement difficult. Without *freedom of movement*, members fall back on a one-person, one-task mentality. Production can decline and teamwork fail. Therefore, obstructions to movement within the cell should be minimal.

In a work cell, visual control of inventory and resources has many advantages. It is simple and accurate. Every team member has the same information.

Each member can take appropriate action to correct imbalances or variations. Visual control takes many forms—kanban boards, queues, and personnel locations. Team members can see delays and unbalanced work, can anticipate upcoming jobs, and can identify and correct quality problems. Such control requires clear lines of sight. It requires mechanisms such as kanban flags, warning lights, and queues. These should be built into the work cell design.

Visual communications build trust. Team members should be

able to see each other and all operations. A clear view of the total operation gives every member the means for controlling the overall process. Partitions, walls, and large equipment interfere with both teamwork and process control.

Ongoing interaction is integral to teamwork. Members must be able to converse during their work. This calls for an area small enough for effective conversation. It also requires a reasonable ambient noise level. While hand signals and electronic communications are partial substitutes, face-to-face conversation is far better.

Effective teams use bulletin boards for many purposes. Bulletin boards help a team communicate with itself, with other shifts, and with the outside world. Prominent space for a bulletin board should be provided.

The A-Team designs a space plan

From the process chart in figure 4.14 and the work cell operations plan in figure 4.18, the team identified ten SPUs. These are on the configuration diagram. Using judgment and consensus, the team members developed the appropriate affinities, most of which resulted from material flow or shared jobs.

Following the procedure for Task 04.04, they developed the configuration diagram in figure 4.21.

Finally, using templates for the equipment and workbenches, the team members prepared a layout. They sized conveyors and queuing areas as they went, fitting these flexible elements into areas between the main workstations. They gradually moved and arranged the SPUs to fit inside the area designated on the micro-space-plan. They allowed clearance for movement of people. They positioned shelving on the perimeter to avoid blocking visual communication.

Selecting the best plan

In Task 04.05, a selection is made from among the micro-layout options. There may be a number of good alternatives for each task and procedure, along with several viable product groups and several good process options. There may be many ways to coordinate, control, and supervise a work cell.

At the micro-level, design iteration is more significant than at the higher levels. Each alternative affects design decisions. The result could be an infinite number of design options. The designer or leader of a design team will make decisions about them as the plan proceeds. Many options will be eliminated. Eventually, two to five fundamentally different designs will be decided upon that gives the evaluation team a good selection and wide choice.

Selection tools for the micro-space-plan

Several sections in Chapter 3 discuss the evaluation of facility options, describing a range of tools to use in evaluating and selecting an option. At the micro-level, the same tools are available.

In practice, micro-level evaluation is simpler and more straightforward than evaluation at higher levels. The range of issues is more limited, the material flow is more obvious, and mistakes are easier to correct. Socio-technical considerations and other intangible issues dominate the selection. These intangibles are less amenable to engineering and financial analysis.

For light assembly or other easy-to-change work cells, the evaluation can be quite informal. The consequences of a less than optimal decision are not severe. In these circumstances, such a decision actually can become a valuable learning experience.

For the above reasons, although material flow analysis or financial methods sometimes are used, selection tools normally are limited to the following:

- simplified cost estimates;
- PNI analysis;
- simulation; and
- weighted factor analysis.

In many situations, a simplified cost estimate and PNI analysis bring strong consensus and further work is unnecessary.

Simulation is valuable in certain situations and has a special place in the evaluation of work cells. It tests proposed changes on existing systems and predicts the performance of new systems. It condenses lengthy processes to a few minutes or slows a rapid process for study. It can experiment where actual physical experimentation is difficult, expensive, dangerous, or impractical.

Computer simulation uses software to create a model for a real or proposed system. A simulation analyst then subjects the model to various inputs and constraints and studies the results. These statistical processes can be repeated hundreds or thousands of times. Almost any statistic can come from a simulation model. Examples are: equipment utilization, throughput time, delivery time, and labor hours.

Computer simulation software uses animation to depict system operation visually. Patients move through a hospital complex. Products advance across the factory floor. People go from task to task. Animation presents simulation results dramatically and effectively. It reveals problems or effects that would be unseen in pure statistical output.

Simulation is most valuable to companies with large investments that are difficult to modify or change. An example is a paper mill where very large machines are on specialized foundations. Once in place, rearrangement is almost impossible. If a process is easy to change, such as a task in a light assembly cell, it may be easier to experiment in real life than to build a simulation.

Simulation cannot model creativity, teamwork, and motivation, which are among the most important benefits of work cells. Specific mechanisms that result from these effects cannot be foretold and therefore cannot be simulated effectively.

Weighted factor analysis can assist with more difficult decisions and is probably the most universal and versatile decision tool. The evaluation procedure is the same as that for Task 03.21 at the macro-level (see Chap. 3).

The A-Team's final micro-space-plan

The space plan in figure 4.21 is not the only one the A Team designed. There were several other good options. Team members used weighted factor analysis to make their decision.

Team members discussed using simulation but decided they could build the actual work cell faster than they could build a simulation model. They also reasoned that if the cell did not work as expected, they could easily change or redesign it. They knew the experience of the work team would be invaluable in any such redesign.

Between the cell and the finished goods warehouse, the kanban system is used. The cell operates with an average of two lots per day; however, it can accommodate special lots at any time. As the team works on improved set ups, the cell may reduce lot sizes or even shift to make-to-order. The team draws material and parts from upstream suppliers and the machine shop using a kanban system.

Each worker sets up his or her own area, changing fixtures, tools, and the parts on their benches. The work team for this work cell is made up of four members, who have various tasks.

"Wilson" subassembles heads, bodies, and coils. He places these subassemblies on a gravity roller conveyor that takes them to Taylor's main assembly station. Wilson uses markings on the roller conveyor to determine the speed of his work. When work builds up on the conveyor, he can switch to restocking or other activities.

"Taylor" subassembles diaphragms one at a time. He then assembles each diaphragm with a body/coil subassembly and the head subassembly. Taylor places his completed work in the queue space between his station and Washington's electrical assembly station. When the queue is full, Taylor assists "Washington."

Washington has the longest series of tasks. He must perform all electrical assembly. He places his finished product in the queue area at his left. In the event that he fills this queue, he can assist "Davis" with electrical testing.

Davis performs the electrical and flow tests and packages the completed pumps. He begins by transferring a unit from Washington's outbound queue to the electrical test fixture. He removes the unit that has just completed the electrical test and places the incoming unit in the test fixture. He records test results on the previous unit while the electrical test goes through an automatic cycle. Davis then places the completed unit in the queue between electrical and flow test and removes one of four units that has been through the flow test. He replaces this unit with the upcoming unit in the flow test queue.

Davis transfers the unit he has just removed from the flow test to the packing table. He packs the unit and places it on an outbound pallet. Since Davis's work does not fill the total design cycle time, he also assists Washington with maintaining the proper stock of electrical parts.

At intervals, a fork truck arrives. The driver removes any completed outbound pallets and leaves kanban cards. These cards become the basis for the next day's schedule. They tell the team what models were removed from finished-goods storage the previous day. The work team arranges these cards on a kanban board. The information on the board tells members which pump models have low inventory and which have full kanban stocks. They plan their schedule to replenish stocks. They also plan the schedule to minimize set-up time by grouping similar models.

At the workday's end, the team meets. They review their performance and plan the next day's production. They also check their inbound stocks to ensure that parts will be available. They may have a problem-solving meeting to address quality or other problems.

Conclusion

This chapter demonstrated how to plan space at the micro-level, Level 4. The discussion included important new developments in the workplace—work cells, teams, and operations focus.

This chapter also offers the tools (model project plan, procedures, techniques) to prepare viable micro-space-plans. In the beginning, following these plans and procedures explicitly and pedantically will result in the generation of good designs. With experience, a design team may use them less formally.

The next chapter covers the final level of space planning, workstation design, including how to place tools and equipment; how to optimize the individual's efficiency and effectiveness; and how to minimize Carpal Tunnel Syndrome and similar occupational hazards.

Chapter 5

Workstation Design: The Sub-Micro-Space-Plan

Written by William Nelson

Sub-micro, or workstation, design is the final space planning level. This is where the firm adds value to its product. The process involves examining tasks, operators, and tools; allocating tasks between the operators and machines; and selecting or designing tools and fixtures. During the process, optimum physical arrangements are developed.

Classical workstation design focused on the output of individual workstations with single operators. Today's design requires a broader view. An optimum workstation design considers: productivity, integration, operator comfort, operator variety, and safety.

Productivity is the efficient transformation of materials and data into a product or service. It is the primary reason for a workstation to exist and, as such, is a primary design consideration.

Integration—technical, psychological, and social—fits the workstation into a larger production system. *Technical integration* allows the occupant(s) of the workstation to perform work at a speed and in a manner that fits with prior and subsequent processes. *Psychological integration* allows the workstation to meet the psychological needs of the operator so that he or she has some control over his or her effort. It gives meaning to the work. *Social integration* in the workstation design allows the operator to mesh well with other people in the production system.

Operator comfort is important. Comfortable operators are inclined to work better. They are more likely to ensure quality output and are more likely to remain in their jobs. Perhaps more importantly, comfortable workstations are less likely to damage the human body with cumulative traumas such as carpal tunnel syndrome.

Most workstations rarely accommodate operators as individuals. Yet, the workplace is awash with a wide mix of diverse individuals, genders, and races. Sound design based on operator variety provides easy adjustments to accommodate each operator—shift to shift and task to task. This is the designer's challenge.

The contemporary regulatory climate demands more attention to safety and health. In addition, the legal and insurance system can impose severe fines on employers that ignore the safety and health of workers. Cumulative trauma disorder (CTD) has become one of the most important areas of occupational health. Preventing CTD is often an important factor in workstation design.

Integration, comfort, variety, and safety are often seen as incompatible with productivity. However, productivity rarely needs to be sacrificed to attain these other design goals. A safe, comfortable workstation that fits well into the larger production system is usually highly productive.

In this chapter, several areas that provide the necessary background knowledge for workstation design will be reviewed. These are: allocation of functions, motion economy, ergonomics, and worker selection. This knowledge will then be applied in a structured and systematic way. A model project plan similar to the plans in other chapters is provided, and an example from Diamond Products illustrates the application.

The information in this chapter demonstrates how to design workstations that optimize productivity, integrate with one another, and improve the work experience. Such workstations contribute to the overall goal—ensuring a productive, comfortable, safe, and healthy workplace.

Allocation of functions

Allocation of functions divides work between people and machines. It determines, to a large extent, the quality of the operator's work experience. A well-thought-out allocation optimizes the interaction of people and machine elements.

In space planning, allocation occurs at several levels. During the macro-space-planning, a process is selected that implicitly allocates many functions. During the work cell design (Level 4), the process is refined and more explicit allocations are made. At the workstation design level, space planners should review these previous allocations.

In a human-machine system, one or more equipment displays show the operator the internal equipment status. He or she processes the information and makes decisions. Using motor responses, the operator alters control settings to change the machine. Continuing observation allows the determination of the effect of altering controls. Figure 5.1 illustrates.

Person-machine systems are pervasive in everyday life. For example, the driver of an automobile depresses the accelerator. He or she observes the speedometer display to ascertain proper speed. If the speed needs adjusting, the driver modifies the depression of the accelerator. Achievement of system goals necessitates the driver's attention. A lapse may result in flashing red or blue lights in the rearview display. These lights signify a failure to achieve at least one system goal—maintenance of speed at or below the posted limit.

In the above example, the system designers allocated speed control to the person. Cruise control reassigns this function under favorable conditions, lessening the driver's workload and achieving another system goal—avoiding speeding citations.

An allocation has three objectives:

- achieve system goals;
- give the worker a coherent set of functions; and
- provide the worker with a reasonable workload.

Some designers tend to allocate functions to people whenever a mechanized solution is not readily apparent. Sometimes, what may seem to be an over-allocation of work to a person in the end provides variety and job enrichment. The variety may prevent injuries from repetitive motion and reduce the cost of workers' compensation insurance and claims. This is an example of an apparently less-than-optimum local system actually bringing increased performance to a larger business system.

Other designers and managers have an inordinate faith in automation. They inappropriately allocate functions to machines; that is, they over-automate.

Automation, or predominantly mechanical allocation, has many benefits and many drawbacks. NASA's industry workshop on automation reported by Boehm-Davis, Curry, Wiener, and Harrison (1983) identified five problems associated with automation:

1. Newly automated systems seldom provide all anticipated benefits.

2. Failure of automated equipment reduces credibility. Operators may not rely on equipment they do not trust.

3. Automation often increases the need for training. The user must learn to operate the equipment in both automated and manual

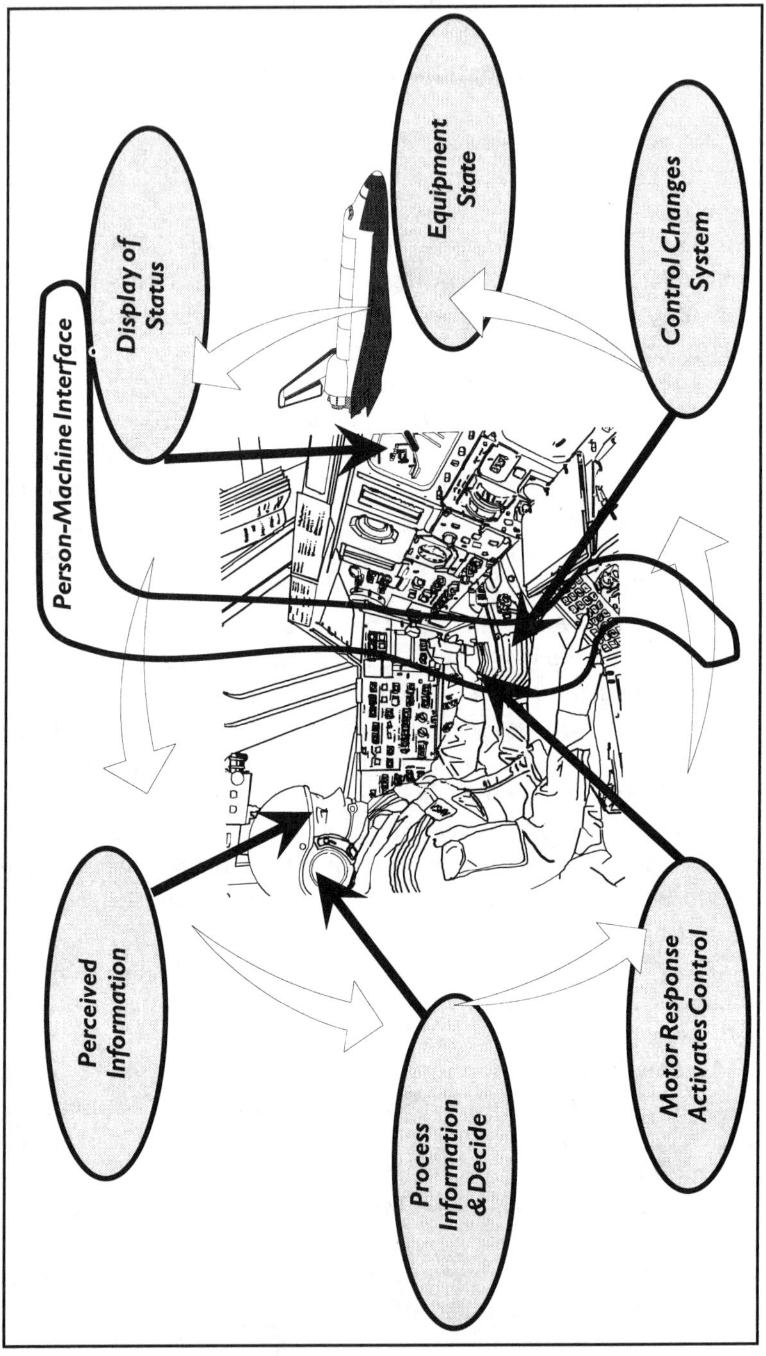

Figure 5.1 - An Example of a Person-Machine System

modes because the equipment may fail at some later date. Increased complexity, common to automated systems, also may increase the training time.

4. System designers seldom anticipate problems created by the automation. They have focused mainly on the benefits of the new system.

5. Automation may transform operators into monitors rather than system controllers. As a result, the operator may not be prepared to take control suddenly if the system fails.

A wide range of allocation options should be considered. This prevents the designer from latching onto an initial concept and ignoring others.

Motion economy

Motion economy helps achieve productivity and reduce CTD. It shortens the human time and effort required to accomplish a task. The principles of motion economy (fig. 5.2) list steps and procedures that simplify and improve manual work.

The fourth principle, *body segment class*, classifies movement with body joints. Each movement after class 1 involves body parts from the previous class(es), and more of the body participates in the motion. It is desirable to accomplish tasks with the lowest possible motion class. The best way to do this is to place everything near the operator. In addition, items should be close together, lightweight, and easily positioned at the end of the motion.

To apply the principles of motion economy, iteration or synthesis may be used. Iteration involves examining an existing or prototype workstation, analyzing the motions, identifying and implementing improvements, and reanalyzing the work. Several repetitions may be necessary.

Synthesis uses predetermined time standard systems such as methods-time-measurement (MTM) and avoids prototyping. These systems refer activities to a standard set of well-defined elemental motions and associated execution times.

The concept of motion economy has limitations. Both iterative and synthetic methods assume workers are not fatigued. Performance can abate over time due to an activity's demands. Motion economy also does not account for physical limitations or differences in operators. Moreover, a movement that appears ineffective from a motion economy perspective actually may prevent fatigue and possible injury from static posture loading. To overcome these limitations, ergonomic principles should be applied.

1.0 Use of Human Body

- The two hands should begin as well as complete their motions at the same time.
- The two hands should not be idle at the same time except during rest periods.
- Motions of the arms should be made in opposite and symmetrical directions and should be made simultaneously
- Hand motions should be confined to the lowest classification with which it is possible to perform the work satisfactorily
- Momentum should be employed to assist the worker whenever possible, and it should be reduced to a minimum if it must be overcome by muscular effort.
- Smooth continuous motions of the hands are preferable to zigzag motions or straight-line motions involving sudden and sharp changes in direction.
- Ballistic movements are faster, easier, and more accurate than restricted(fixation) or "controlled" movements.
- Rhythm is essential to the smooth and automatic performance of an operation, and the work should be arranged to permit easy and natural rhythm wherever possible.

2.0 Arrangement of Work Place

- There should be a definite and fixed place for all tools and materials.
- Tools, materials, and controls should be located close in and directly in front of the operator.
- Gravity feedbins and containers should be used whenever possible.
- Drop delivers should be used whenever possible.
- Materials and tools should be located to permit the best sequence of motions.
- Provisions should be made for adequate conditions for seeing. Good illumination is the first requirement for satisfactory visual perception.
- The height of the workplace and the chair should preferably be arranged so that alternate sitting and standing at work are easily possible.
- A chair of the type and height to permit good posture should be provided for every worker.

3. Design of Tools and Equipment

- The hands should be relieved of all work that can be done more advantageously by a jig, a fixture, or a foot operated device.
- Two or more tools should be combined whenever possible.
- Tools and materials should be prepositioned whenever possible.
- Where each finger performs some specific movement, such as in typewriting, the load should be distributed in accordance with the inherent capacities of the fingers.
- Handles, such as those used on cranks and large screwdrivers, should permit as much of the surface of the hand to come in contact with the handle as possible. This is particularly true when considerable force is exerted in using the handle. For light assembly work the screwdriver handle should be so shaped that it is smaller at the bottom than at the top.
- Levers, crossbars, and handwheels should be located in such positions that the operator can manipulate them with the least change in body position and with the greatest mechanical advantage.

Body Segment Classes

Class	Body Joint	Body Segments
1	Knuckle	Fingers
2	Wrist	Hand, Fingers
3	Elbow	Forearm, Hand, Fingers
4	Shoulder	Upper Arm, Forearm, Hand, Fingers

Figure 5.2 - Principles of Motion Economy

Ergonomics

Ergonomics is the study of work as it relates to the human body and its limits. The usual goal is maximizing output without physically harming the operator. To achieve this goal, designers adapt tasks and the workstation to individuals, not vice versa. *Physiology*, *biomechanics*, and *anthropometrics* are the areas of ergonomics most useful to the designer of workstations.

Physiology

In some respects, the body is analogous to an automobile. In the human machine, muscles are both cylinders and pistons, and bones and joints are the gears. The muscles oxidize nutrients (fuel) and give up energy, while generating metabolic byproducts (waste). Physiology studies this process.

Two categories of physiological demands usually are relevant during work: static and dynamic. Most activities combine static and dynamic postures. While some muscle groups have a static posture, others have dynamic postures.

Static work occurs when the body is in a stationary position for an extended period. The musculoskeletal system is unsuited for prolonged static work because the body cannot supply fresh nutrients to the stressed tissues. In addition, waste products remain at the stressed site. Muscles and tendons can become inflamed. Even at static loads as low as 30 percent of maximum strength, fatigue develops rapidly.

In dynamic work, the body is in motion. Nutrients and waste products move to and from the muscles. Consequently, the muscles can work for extended periods if the maximum load on the body is significantly less than the maximum static capability. Endurance usually limits dynamic work when loads are not extreme.

Usually, tasks should not require operators to exert more than 30 percent of their maximum muscle force in a prolonged or repetitive way. All muscular exertions beyond 50 percent of the maximum level should be avoided.

Biomechanics

Biomechanics is the study of mechanical forces in human movement, including the interaction between individuals and their physical environment. Biomechanical principles primarily are used to minimize damage to muscles, joints, and tissues. This damage may come from a one-time force, such as lifting an object that is too heavy or moving an object from an awkward position. Damage also can come from an accumulation of small, repetitive forces—CTD.

There are three actions in the work place that can cause damage: extreme joint movements, excessive force, and highly repetitive movements.

Extreme joint movements, such as bending the wrist, amplify the forces placed on the joint. They may prevent the operator from applying maximum force and increase damage that results from the force that the operator does use.

Excessive force used for lifting, squeezing, or pushing is a primary cause of injury. Often, such excessive force combines with repetition or extreme joint movement. A need for excessive force does not always arise from a workstation's design: it may come from poor maintenance of tools and equipment or from operator practice or ignorance.

Repetition increases the damaging effect of muscular forces. The more frequent and constant the repetition, the greater the damage. Even the small force of operating a keyboard can produce the debilitating effects of carpal tunnel syndrome.

Figure 5.3 recommends methods for reducing and avoiding these actions.

Anthropometry

Anthropometry studies the dimensions, weights, and strengths of human body segments. Anthropometric data aid in designing the workstation to the operator's dimensions.

Anthropometry uses static (structural) and dynamic (functional)

1.0 Extreme Joint Movements

- Alter the tool or control- bend the tool or handle instead of the wrist
- Move the part- rotate the part in front of the worker to keep the wrist straight
- Move the Worker- change the worker's position relative to the part
- Avoid reaching above shoulder level
- Avoid reaching behind the body
- Keep elbows close to the sides
- Place the work area about 2"-4" below the elbow when standing or seated in an erect posture
- For precise or delicate tasks, place the worksurface 4"-8" above elbow height.
- For heavy manual assembly, place the worksurface 4"-5" below elbow height.
- Start your design from the working point where the hands spend most of their time

2.0 Excessive Force

- Keeping cutting edges sharp and tools well maintained.
- Spread Force- Alternate hands, use levers instead of buttons.
- Increase Mechanical Advantage- Use stronger muscle groups and long handles
- Use jigs and fixtures whenever possible
- Select gloves carefully. They can reduce grip strength up to 15%.

3.0 Repetitive Movement

- Task Enlargement- Give workers larger and more varied tasks and increase cycle time.
- Mechanization- Use special tools with ratchets or power drivers.
- Automation- Allocate Repetitive to machines.
- Give the operator a neutral posture.
- Allow variation of method to prevent a static posture for extended periods.
- Permit several working positions
- Re-sequence jobs to reduce repetition

4.0 Physiology

- Allow self pacing of work when possible
- Allow frequent rest for most active muscles
- Start new employees at a slower rate

5.0 Hand Tools

- Provide handles
- Design For minimum muscular effort
- Power with motors more than with muscles
- Bend the tool and not the wrist
- Keep the effective weight of the tool low
- Align the tool center of gravity and the center of the grasping hand
- Use pistol grips when the tool axis is horizontal
- Use straight grips when the tool axis is vertical
- Use trigger levers rather than trigger buttons.
- Design special use tools if needed
- Design tools for use by either hand
- Use A Minimum handle length Of 4"
- Use proper size grips which accommodate different size hands
- Spring load pliers and scissors
- Use non-porous, non-slip, and non-conductive grips
- Tools should weigh less than 9 lbs
- Suspend heavy or awkward tools

6.0 Position

Use a standing position when:

- Knee Clearance Is Unavailable
- The Operator Lifts More than 10 Pounds
- There are high, low, or extended reaches
- The Operator Exerts Downward forces (wrapping and packing)
- The Operator Needs Mobility

Use A sit/stand position when:

- Repetitive operations have frequent reaches beyond zones 1
- The Operator performs both sitting and standing tasks
- The Task Requires Prolonged Static Effort

Use a sitting position when:

- Items for a repetitive, short cycle are in seated work space.
- Hands work less than 6" above the surface
- Large force is not required
- Handling weight is less than 10 lbs
- The task is fine assembly or writing
- The operator needs body stability and equilibrium
- The task requires precise foot control
- Operator has extended time in a fixed position

Figure 5.3 - Principles of Ergonomics

measurements. Static dimensions are the measurements between specific anatomical landmarks. Examples include stature, arm length, and shoulder breadth. Dynamic dimensions relate to functional movement. Both measurements are important. For example, static data might represent a person's reach when his or her shoulders and trunk are stationary. Dynamic measurement would add the length of that person's reach from extended shoulders and trunk. In many cases, static dimensions may be adapted to a specific problem. Dynamic dimensions, on the other hand, are better representations of reality.

The basis of the data should always be kept in mind. Does it include clothing and shoes? What specific population was included? What about those with disabilities? Designers should use intuition and experience as well as raw data.

Classical design considered the average male the standard. Today's designers may need to consider everyone from the smallest female up to the largest male. Different ethnic populations should also be included. The range of anthropometric data is great. Designers must determine which criteria apply to the problem at hand.

Dimensions for several groups of males and females are given in table 5.1. How they were taken is shown in figure 5.4. Several of these data groups were based on U.S. Air Force flying personnel; therefore, very large and very small people probably are not represented.

There are four constraints of ergonomic anthropometry: *clearance*, *reach*, *posture*, and *strength*. Clearance provides adequate clearance for those in the workstation. This includes headroom, elbow room, leg room, and handle space on a hand tool.

Clearances are designed for the maximum. For example, if the tallest person in a group has clearance, shorter people will have it also. Figure 5.5 shows several typical workplace clearances. In most of the examples, three values in inches are given. The first is the minimum normal clearance, the second is the clearance with normal clothing, and the third is the clearance required with heavy winter clothing.

Reaching is a common workplace activity. Workers reach for parts, tools, and controls. Reach constraints determine the maximum acceptable distance of the item—an example of designing for the minimum. If the individual with the shortest reach can grasp, those with longer reach can also grasp. The distances in figure 5.4 include static reaches. Figure 5.6 shows reach zones that vary according to difficulty.

Postural constraints often are difficult to identify because they depend on the size of the person, his or her position, and the equipment dimensions. The height of a work surface, for example, depends on the height of the worker and the height of the worker's chair. With postural

TABLE 5.1

ID	ID	Description	Med	SD	Min	Max	GP	Med	SD	Min	Max	GP
1	A	Height	69.1	2.44	59.5	77.6	a	63.2	2.48	55	73	d
3	B	Shoulder Breadth	17.9	0.91	14.6	22.8	a	13.4	1.22	8.7	19.3	d
5	C	Hip Breadth (Seat)	15.3	1.11	12	21.3	b	14.6	1.04	12.1	20.6	c
5	C	Hip Breadth (Stand)	13.2	0.73	8.3	15.8	a	15	1.03	11.8	18.9	e
5	C	Hip Breadth (Seat)	13.9	0.87	11.4	18.1	a					
19	D	Foot Breadth	3.8	0.19	3.2	4.7	a					
11	E	Arm Span	70.8	2.94	58.3	82.3	a					
10	F	Arm Reach	34.6	1.65	27.6	39.8	d	31.8	1.29	28.3	35.4	e
4	G	Chest Depth	9	0.75	6.7	13	a					
20	H	Hand Breadth	3.5	0.16	3	4.1	a					
9	I	Forearm-Hand	18.9	0.81	15.4	22.1						
13	J	Buttock-Knee	23.6	1.06	18.5	27.6	a	22.6	0.96	19.7	26.7	e
14	K	Seat Length	18.9	0.96	15.4	23.1	b	18.2	1.04	15.3	22.2	c
15	L	Sitting Knee Hgt	21.7	0.99	17.3	24.8	a	17.2	1.07			d
18	M	Foot Length	10.5	0.45	8.9	12.2	a	9.6	0.4	8.9	10.9	e
2	N	Erect Sit Height*	36	1.29	29.9	40.2	a	34.1	1.02	30.7	34.4	e
6	P	Shoulder Hgt (Seat)	23.3	1.14	18.9	27.2	a	24.6	3.02			d
16	Q	Seat Height	19	0.89	15.6	22	b	18.1	0.89	15.4	20.6	c

Notes:
Source: McCormick, 1964
*Normal Sitting Height is about 2" Less Than Erect Height
Group "c" Data Are Medians, Groups "d" and "e" Data Are Means

Groups:
a — 4000 Air Force Flying Personnel (Herzberg, Daniels and Churchill)
b — 1959 Civilian Males (Hooten and Staff)
c — 1908 Civilian Females (Hooten and Staff)
d —10,042 Civilian Females (O'Brien and Shelton)
e — 447 Female Pilots (Randall, Damon, Benton and Pratt)

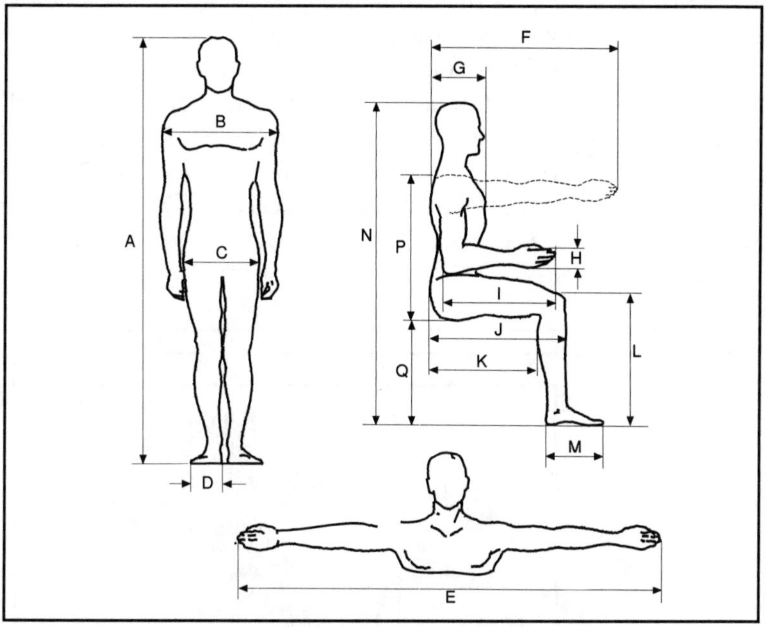

Figure 5.4 - Typical Anthropometric Dimensions

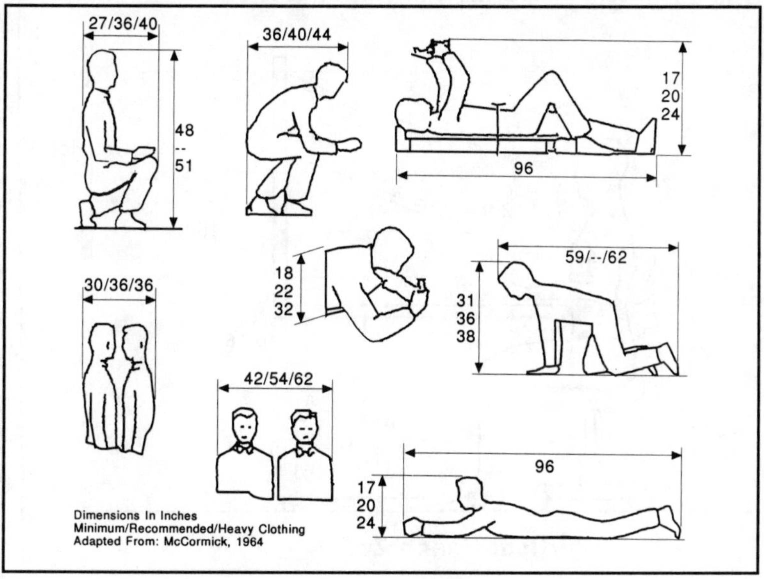

Figure 5.5 - Typical Work Clearance

constraints, designing for the extremes is the common methodology. Figure 5.6 shows several postural dimensions.

Strength requires focus on the minimum in one situation and the maximum in others. It depends on the direction of force and the position of the operator, as well as the strength of the individual operator. Experience and a conservative approach will help to ensure that the operator's strength limits are not exceeded. For a straight lift under ideal conditions, an operator should never lift a weight of more than fifty-one pounds.

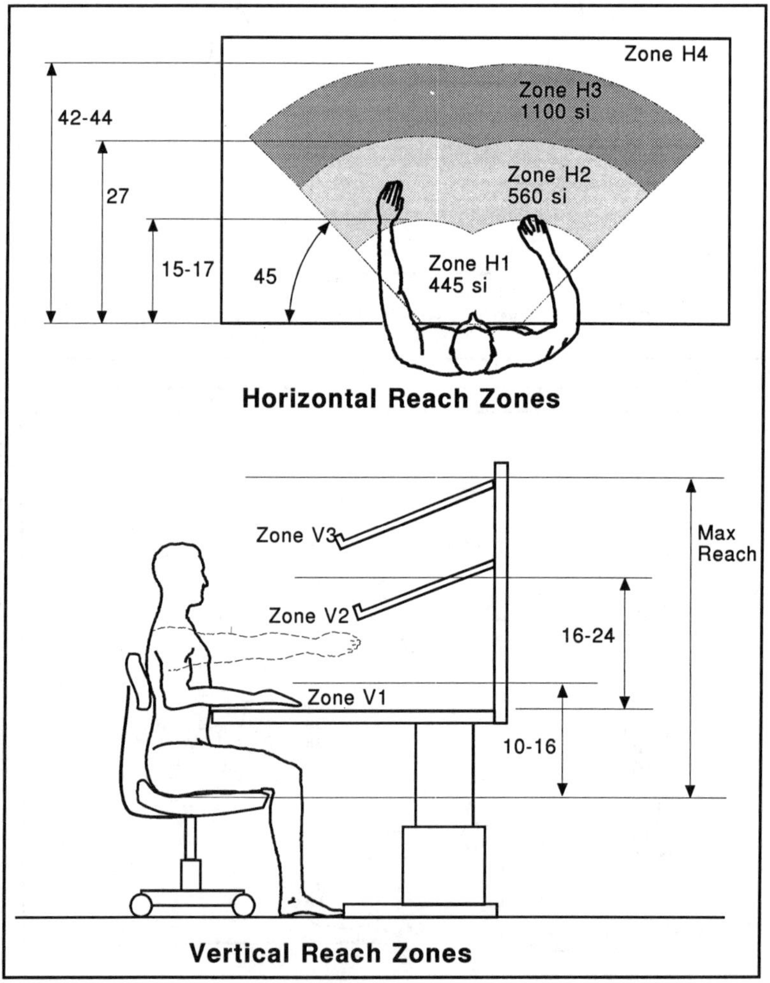

Figure 5.6 - Reach Zones

Special topics

Several specific topics arise repeatedly in the ergonomics of workstation design. Among these are *hand tools*, *material handling*, *work positions*, and *seating*. In addition to the discussion below, figure 5.3 contains principles that apply to these areas.

Hand tool selection, design, and use are important elements of many workstation designs. The use of the correct hand tools contributes to productivity and quality and can prevent CTD and other injuries.

Almost every workplace requires some form of material handling. Injuries caused by maneuvering materials manually cost industries well over $15 billion annually in direct expenses. The indirect costs are an estimated at $60 to $75 billion. Productivity losses from poor handling methods are probably even greater. Any tasks that require manual material handling should be carefully reviewed. Refer to the National Institute for Occupational Safety and Health's (NIOSH) Revised Lifting Equation (1991).

Operators may sit, stand, or both. Work positions that combine sitting and standing permit operators to shift their postures. This helps reduce muscle fatigue from prolonged static effort. Figure 5.7 summarizes the relationship between several workplace variables and preferred seating positions.

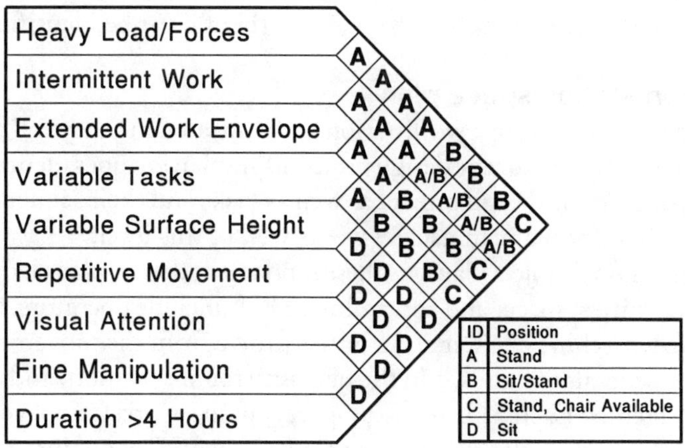

Figure 5.7 - Work Positions

Researchers have studied seats extensively, and most designers have broad, practical experience with seats. In general, a seat that is comfortable for the task and positions the arms properly is ergonomically satisfactory.

Only the most important ergonomic factors in workstation design have been covered. Lighting, vibration, temperature, noise, and shift work also affect the design. For further information, see Salvendy's *Handbook of Human Factors* and McCormick's *Human Factors In Engineering and Design.*

Worker selection

The physical demands of every job differ, as do the physical, mental, and temperamental characteristics of each person that performs a job. For optimum performance, these characteristics and capabilities should match.

An employer should identify characteristics required for each job and identify minimum criteria for those who might be selected for it. This will ensure an appropriate match. In some cases, the identification of characteristics may result in a job redesign to accommodate a larger part of the available candidates. At minimum, employers should ensure that the people in a particular job do not incur physical harm because of a mismatch between job requirements and worker characteristics.

The Americans with Disabilities Act (ADA) is another reason to perform this analysis. This act requires employers to accommodate, within reason, people with disabilities who can perform the essential functions of a job. To do so means that the essential functions and the worker characteristics needed to perform them must be identified.

The workstation space plan

The tasks for preparing a workstation are illustrated in figure 5.8. Task 05.01 sets out a plan for the design project. The plan includes identifying participants in the design project as well as tasks and their sequence. A simple and informal plan usually works well at this level.

The model project plan uses these fundamental space plan elements: SPUs, affinities, space, and constraints. SPUs include operators, tools, parts, and machine elements. Affinities evolve from the movement of parts, movement of hands, and the necessity for observation, and other indirect factors. Each SPU requires space, but in a workstation, vertical space is more important than it is at other levels. Constraints have a different quality. They now include an operator's dimensional and physiological limits and, possibly, regulatory limits. Process constraints may play an important role.

In Task 05.02, designers bring together the necessary information.

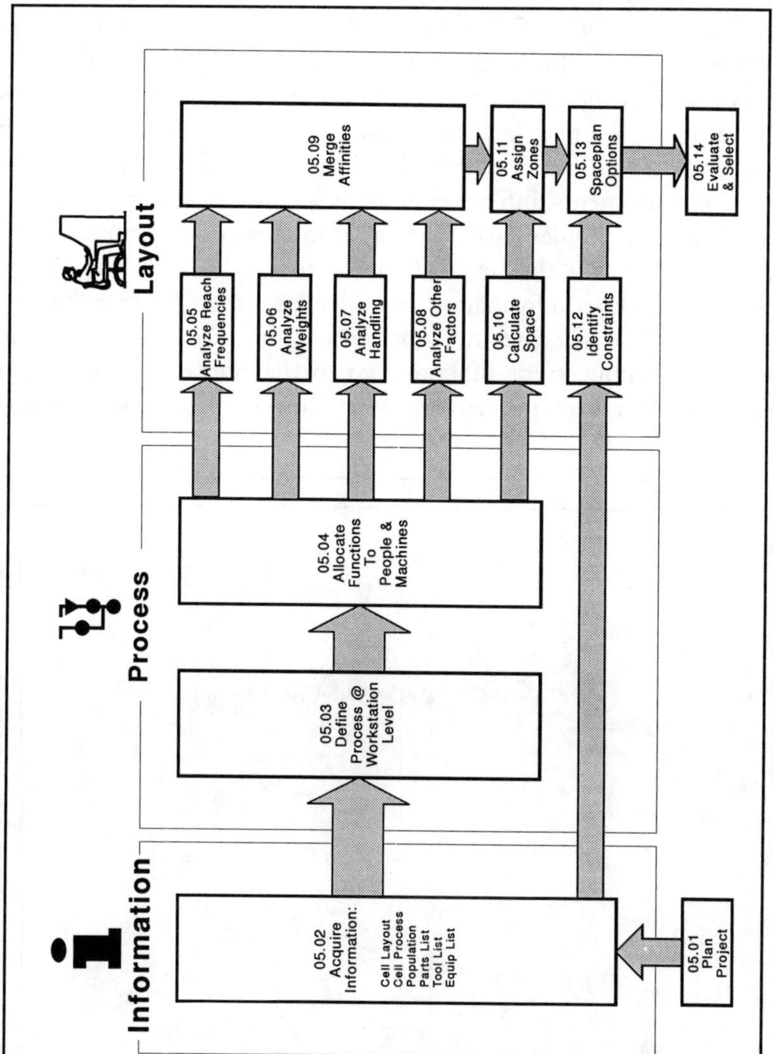

Figure 5.8 - Workstation Model Project

This includes a layout of the area, a process chart for the overall process, information on the people who will work at the station, parts lists, tool lists, and equipment lists. If the workstation design follows a cell design, much of the necessary information may already exist. If not, it will have to be obtained or generated. Chapter 4 shows how to acquire this information.

Next, a detailed process chart for workstation activity is created (Task 05.03). Figure 5.9 is the workstation process for the head subassembly of the diaphragm pump described in Chapter 4. The cell-level chart identified this subassembly as a single operation. Now, it is exploded into the next level of detail, and each component and each operation that uses a separate tool or instrument are documented. In service environments, information items go into the chart.

Table 5.2 includes a list of parts and tools for the head subassembly. This spreadsheet is the heart of the layout analysis. As the design proceeds, the other information and calculations will be described.

Task 05.04 allocates functions to people or machines. For the head subassembly, most functions go to the operator. The design volume is eighty-eight pumps per day or about eleven pumps each hour.

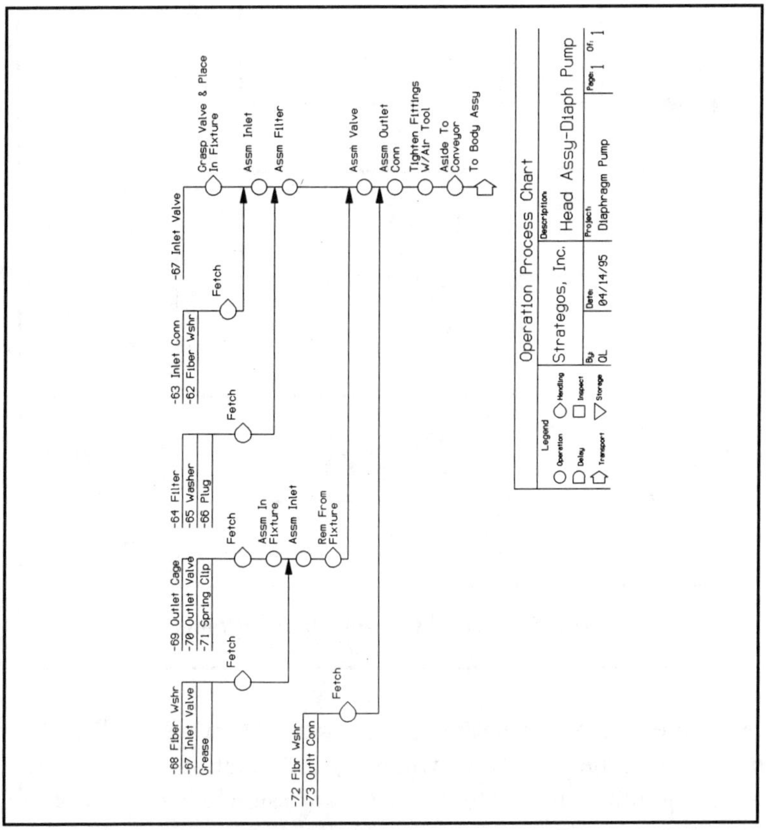

Figure 5.9 - Operation Process Chart

(A) ITEM ID	(B) ITEM Description	(C) Qty	(D) UN/Day	(E) CONTAINER COD	(F) CONTAINER Len	(G) CONTAINER Wid	(H) CONTAINER Hgt	(I) REACH Frq	(J) REACH Aff Rat	(K) WEIGHT Wgt	(L) WEIGHT Aff Rat	(M) Hdl Rtg	(N) Otr Rtg	(O) Tot Scr	(P) Total Aff	(Q) Zone Assgn	(R) AREAS Hor	(S) AREAS Face	(T) ZONE Sub Totals
29	ASSEMBLY FIXTURE	1	88		14.0	12.0	2.5	1320	4	0.00	1	0	4	N/A	4+	H1V1	202	36	202
28	GREASE TUBE	1	88		10.0	3.0	3.0	88	4	0.20	1	2	0	2.8	4	H2V1	36	11	
27	AIR WRENCH	1	88		N/A	N/A	N/A	88	4	0.30	1	1	0	2.7	4	H2V1	0	0	
7	8B10-67 INLET VALVE	1	68	BN2	6.0	4.0	7.5	68	3	0.07	1	4	0	2.6	3	H2V1	29	36	
10	8B10-70 OUTLET VALVE	1	68	BN2	6.0	4.0	7.5	68	3	0.07	1	4	0	2.6	3	H2V1	29	36	
11	8B10-71 SPRING CLIP	1	68	BN2	6.0	4.0	7.5	68	3	0.01	0	4	0	2.6	3	H2V1	29	36	
1	8B10-61 HEAD-BODY	1	68	BK4	20.0	13.0	6.0	68	3	2.80	4	1	0	2.6	3	H2V1	312	94	434
9	8B10-69 OUT VLVE CAG	1	68	BN2	6.0	4.0	7.5	68	3	0.02	0	3	0	2.4	3	H2V2	29	36	
2	8B10-62 FIBER WASHER	1	68	BN1	12.0	6.0	7.5	68	3	0.01	0	3	0	2.4	3	H2V2	86	54	
5	8B10-65 WASHER	1	68	BN1	12.0	6.0	7.5	68	3	0.01	0	3	0	2.4	3	H2V2	86	54	
8	8B10-68 THIN FBR WSH	1	68	BN2	6.0	4.0	7.5	68	3	0.01	0	3	0	2.4	3	H2V2	29	36	
12	8B10-72 MED FB WSHR	1	68	BN2	6.0	4.0	7.5	68	3	0.01	0	3	0	2.4	3	H2V2	29	36	
4	8B10-64 FILTER	1	68	BN1	12.0	6.0	7.5	68	3	0.08	1	2	0	2.2	3	H2V2	86	54	
20	8B10-67 INLET VALVE	1	20	BN2	6.0	4.0	7.5	20	2	0.07	1	4	0	2.0	2	H2V2	29	36	
23	9B10-70 OUTLET VALVE	1	20	BN2	6.0	4.0	7.5	20	2	0.07	1	4	0	2.0	2	H2V2	29	36	
24	9B10-71 SPRING CLIP	1	20	BN2	6.0	4.0	7.5	20	2	0.01	0	4	0	2.0	2	H2V2	29	36	432
14	9B10-61 HEAD BODY	1	20	BK4	20.0	13.0	6.0	20	2	2.80	4	1	0	2.0	2	HV31	312	94	
3	8B10-63 INLET FTG	1	68	BN1	12.0	6.0	7.5	68	3	0.15	1	0	0	1.8	2	H3V1	86	54	
13	8B10-73 OUTLET FTG	1	68	BN1	12.0	6.0	7.5	68	3	0.15	1	0	0	1.8	2	H3V1	86	54	
6	8B10-66 PLUG	1	68	BN1	12.0	6.0	7.5	68	3	0.10	1	0	0	1.8	2	H3V1	86	54	
22	9B10-69 OUT VLVE CAG	1	20	BN2	6.0	4.0	7.5	20	2	0.02	0	3	0	1.8	2	H3V1	29	36	
15	9B10-62 FIBER WASHER	1	20	BN2	6.0	4.0	7.5	20	2	0.01	0	3	0	1.8	2	H3V1	29	36	
18	9B10-65 WASHER	1	20	BN2	6.0	4.0	7.5	20	2	0.01	0	3	0	1.8	2	H3V1	29	36	
21	9B10-68 THIN FBR WSH	1	20	BN2	6.0	4.0	7.5	20	2	0.01	0	3	0	1.8	2	H3V1	29	36	
25	9B10-72 MED FBR WSHR	1	20	BN2	6.0	4.0	7.5	20	2	0.01	0	3	0	1.8	2	H3V1	29	36	
17	9B10-64 FILTER	1	20	BN2	6.0	4.0	7.5	20	2	0.08	1	2	0	1.6	2	H3V1	29	36	
16	B910-63 INLET FTG	1	20	BN1	12.0	6.0	7.5	20	2	0.15	1	0	0	1,2	1	H3V1	86	54	
26	9B10-73 OUTLET FTG	1	20	BN1	12.0	6.0	7.5	20	2	0.15	1	0	0	1.2	1	H3V1	86	54	
19	9B10-66 PLUG	1	20	BN1	12.0	6.0	7.5	20	2	0.10	1	0	0	1.2	1	H3V1	86	54	1003
														2.1	2		1870	1224	1870

RATING WEIGHTS		AVAILABLE SPACE (SI):	
REACH FREQ:	0.6	H1V1:	445
WEIGHT:	0.2	H2V1:	560
HANDLING:	0.2	H2V2	560
OTHER:	0	H3V1:	1132
TOTAL:	1		2697

TABLE 5.2

Moreover, two sizes make up this volume. In the designer's judgment, automation was impractical at this volume. The only machine assignment is the tightening of fittings. The operator uses an air wrench to do this.

Tasks 05.05 through 05.08 address affinities. Most affinities arise from reaching, grasping, and moving. The operator may interact in this manner with parts, tools, or controls. A reach, followed by a grasp, is one of the most common workplace activities. Operators reach and grasp parts, tools and controls. Frequency, handling difficulty, and weight affect the reach affinity.

Affinities also arise from other factors. The operator may need to see objects or displays, so they should be within the operator's visual field. Some items are more important than others. For example, an emergency cutoff control is used only infrequently but is very important. Developing affinities also involves: the amount of accuracy required, duration of use, safety, the amount of force required, and operator preferences.

The various tools, parts, displays, and controls compete for locations close to the operator and within his or her optimum visual and reach zones. These zones have limits.

Figure 5.6 shows the horizontal and vertical zones for a typical operator. These are designated with an "H" or "V" for horizontal or vertical. Designers then can assign items to both a horizontal and vertical zone according to their relative affinity needs and the space available.

Horizontal affinity zone 1 (H1) is closest to the operator and can be reached comfortably when elbows are on the work surface. The zone arch is about 15 to 17 inches from the surface edge. A line extending 45 degrees from the shoulders bounds the zone laterally. This zone is the most comfortable, offering the least stress and quickest access time. It should be reserved for the tasks and items with the highest priority.

Horizontal affinity zone 2 (H2) is the area an operator can comfortably reach by extending an arm with his or her trunk stationary. The zone arc is about 27 inches from the work surface edge. It ends when the arm is at 45 degrees above the horizontal plane. This zone requires minimal access time. However, extensive time in this position leads to rapid fatigue and upper musculoskeletal stress. Use this zone for light objects that the operator can grasp easily before returning to zone H1.

Horizontal affinity zone 3 (H3) is reached by extending an arm with full trunk flexion. The arc is about 42 to 44 inches from the work surface edge. It ends when the arm reaches 45 degrees above the horizontal plane. This zone has significantly greater access time and corresponding reduction in efficiency, productivity, and effectiveness. Operators must move their heads to see the task, a movement that interrupts concentration. This zone should be used for infrequent reaches with low priority.

Horizontal affinity zone 4 (H4) requires full body movement and, possibly, one or two side steps. The operator expends considerable energy to reach this zone. Access time is significantly greater than other zones. It should only be used for tasks with the lowest priority.

Vertical affinity zone 1 (V1) extends from the work surface to about 10 to 16 inches above the seat pan. It is about heart level, has the lowest access time, and is the most comfortable. It should be used for tasks with the highest priority.

Vertical affinity zone 2 (V2) begins where zone V1 ends. It extends to the operator's shoulders, about 16 to 24 inches above the work surface. This zone is used for items and tasks that support primary tasks. Access time is minimal, but extensive time in this zone causes early fatigue and musculoskeletal stress.

Vertical affinity zone 3 (V3) extends from the base of zone V2 to the maximum reach level. It may require upper trunk movement as well as head movement, which reduces concentration and increases access time. Using this zone interrupts vision and forces the operator to refocus. It is the least efficient vertical affinity zone and should be reserved for tasks with the lowest priority.

The first task for affinity development concerns reaches and frequency. From the initial information, designers should compile a list of items. In table 5.2, this list is in column B. Next, designers should use P-V information, the process chart, and bills of materials to determine the number of times each day an operator reaches each item. In table 5.2, this is column I. The frequencies derive from the quantity per unit (column C) and the expected daily production (Column D).

Column J is an affinity rating that uses the 0 to 4 weighted numeric scale. The rating is identified on a ranked bar graph, a process similar to that used for rating material flow affinities discussed in chapter 3. To construct this graph, the spreadsheet on column I is sorted in descending order. Column I is plotted on the vertical axis while the identifying number, column A, is plotted on the horizontal axis. The next step involves assigning 0 to 4 or AEIOU to the affinities and recording them in column J of the spreadsheet.

The weight of each object, if significant, adds to the difficulty of a reach. Column K lists the weight carried for each item, which is then rated by the corresponding affinities (column L). Difficulty of a reach-grasp also depends on the shape, size, and delicacy of each item. This affinity is rated in column M. For the diaphragm pump, the valves are particularly small and delicate. Consequently, they carry a handling affinity of 4. Small washers carry a 3 affinity.

For the head subassembly of the diaphragm pump, no other affinity factors were identified. In different situations, this category could include the necessity to see gauges or controls. It could include many factors peculiar to each circumstance. Using the procedures for Task 03.14 outlined in chapter 3, affinities are merged. At the workstation level, this is Task 05.09. Each affinity is assigned a rating weight. Table 5.2 shows these weights at lower left. Each affinity rating is multiplied by the corresponding weight and placed in column O. Next, the values in column O are plotted on a ranked bar graph, and the total affinities are rated and are placed in column P.

Task 05.10 calculates the space for each item in the list. Table 5.2 has dimensional information in columns F through H. Column E shows a container code. For example, the BN1 container is an open-front parts bin with dimensions of 6 inches by 7.5 inches by 12 inches. The spreadsheet formula calculates horizontal space and places it in column R. The face area for each item is shown in column S. The formula adds 20 percent for unusable space and clearances.

Task 05.11 assigns each item an affinity zone. This balances the need for affinity with available space. Starting with the highest value affinities, each item is assigned to the most desirable zone. For the head subassembly, zone H1V1 was reserved for the work fixture and surrounding work space. The lower middle section of table 5.2 shows the approximate space available in each zone. Because the total amount of space needed exceeds the space available in all the V1 zones, the designers use an elevated shelf in zone H2V2 to store some parts.

In this situation, horizontal space dominates. In other situations, such as in an aircraft cockpit, the vertical space might dominate. For this example, affinities between elements other than the operator have been ignored. Such affinities might have importance in special situations. In such cases, designers develop an affinity diagram like that used in Chapters 3 and 4.

Task 05.13 creates space plan options. One option for the head subassembly is in figure 5.10. The designers used figure 5.7 to select a position for the operator. In this case, the operator assembles heads and bodies. The operator also stocks material and occasionally assists other people in the work cell. The sit and stand position offers mobility but also allows rest and variation.

Figure 5.10 shows how the operator can assemble either of the two pump sizes without set-up. This precludes batching. At completion, the parts go on a roller conveyor at the operator's left. The roller conveyor and one-part container are in Zone 4, outside the usual affinity zones. The operator uses these only once for each item. A swivel seat prevents awkward bending for both reaches.

A second roller conveyor feeds parts for the body assembly. This assembly occurs at the workstation to the operator's left. It also carries the completed body assemblies.

The operator's air wrench hangs on a tool balance. An adjustable task light reduces glare. Adjustable height on the bench and seat accommodates all operators. The operator stocks material from the back side of the bench, which is adjacent to aisle and storage areas.

Task 5.14 is the evaluation of the space plan. The evaluation tools from Chapter 3 should be used.

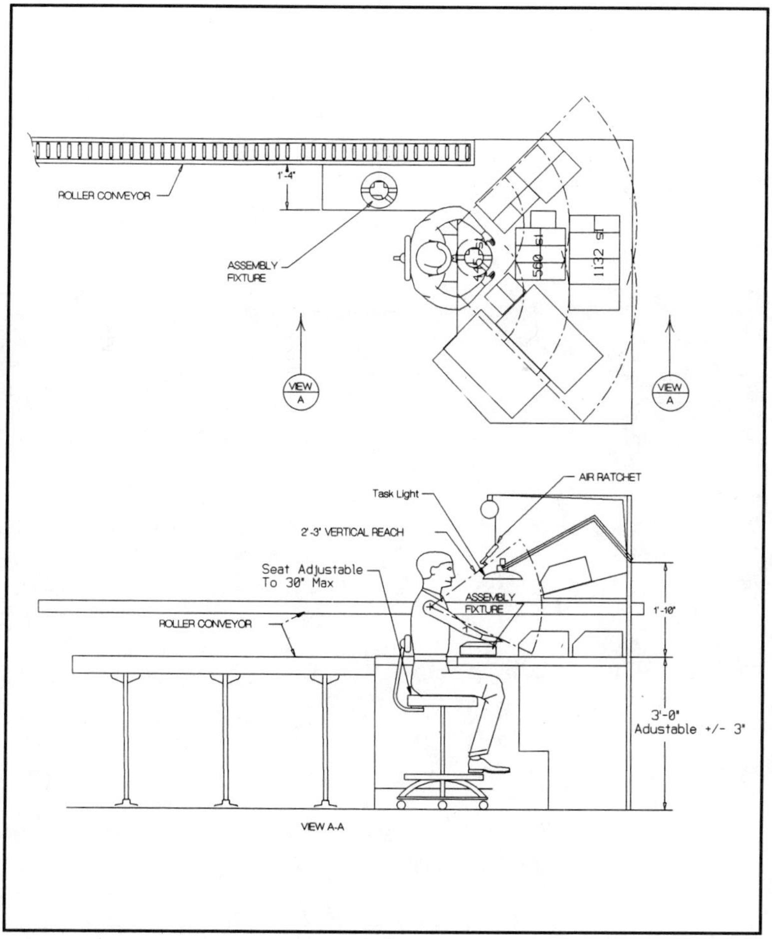

Figure 5.10 - Workstation

Summary

A well-thought-out workstation optimizes productivity both within itself and in the larger production system. It improves the work experience for everyone. It ensures continued health and safety. Such workstations are an essential part of a world-class operations strategy. A structured and systematic approach will ensure consistent and high-quality workstation design.

Chapter 6

Site Planning and Location

Written by Arild Eng Amundsen

This chapter discusses the location, acquisition, and planning of sites. These are Levels 1 and 2 in figure 2.1. At the global level, Level 1, management decides the approximate geographic location for the site(s). Such decisions necessarily include the scope and range of activity on each site, as well as scale issues. They include integration—how the sites relate to each other and the outside world. First, a desirable region or area for the site is located. Management then finds specific site candidates, makes a selection, and negotiates the acquisition.

The supra-level, Level 2, includes planning the location of buildings on the site. Other significant site features such as utility lines, roads, railways, and green areas are also positioned. Zoning and neighborhood issues are examined, and future expansions and development are explored.

Because sites are so permanent and expensive, such decisions are both long-range and strategic. Only top management has the authority and responsibility for such decisions. Industrial engineers can assist management with these decisions by gathering facts, framing the issues, and guiding the debate. They also can compose significantly different and important options and provide the tools for evaluation.

Ideally, site location and planning procedures should occur sequentially and before other facilities planning (see Fig. 2.7). However, this may not happen. Management already may have purchased a site, or an older, existing site may need replanning. Designers may have to rationalize the scope, range, and scale of activities for multiple existing sites. They must integrate and combine compatible products, markets, and activities. These less-than-ideal situations are common.

Site location and planning issues are often intangible, probablistic, or conditional. The tasks are seldom consistent from project to project. The methodologies are inexact and often not amenable to engineering analysis. Most industrial engineers face these tasks only occasionally in their careers. For these reasons, site location and planning require more experience than other levels of space planning. Organizations may wish to engage the services of specialists in this field.

In this book, single-building sites are emphasized. This is the most common situation, particularly for those who are relatively new to facility planning. Large, multi-building sites are far more complex and involve a more complex project structure. They also involve the larger issues of focus, detailed construction, and structural engineering.

Site planning and location can follow the same principles detailed in the previous chapters. A framework of detail levels translates into planning phases. For each phase, model project plans are used to develop a project plan.

Site location and acquisition

An existing firm that simply needs a more suitable site in the current region does not need to be analyzed on a global level. However, a firm planning an additional site or a move to a new geographic region faces strategic and long-term issues. A sound site-location project is not merely a search for cheap labor and low taxes. Increasingly, critical skills in labor and support determine the success of a firm. Geopolitical factors may influence the decision.

The phasing between Levels 1 and 2 may not be clear and sequential. For example, initial site plans may have to be prepared before anyone can decide among final candidates. After the acquisition, more detailed site plans assist the architects and civil engineers as they design the buildings and key features.

The model project plan for Level 1 activity is in figure 6.1. It starts with a preliminary investigation. This information acquisition (Task 01.01) usually takes the form of discussion with top management in marketing, operations, and finance. It may require gross cost estimates to verify that expectations are realistic. The deliverables are project plans for Level 1 and 2 and a report.

Task 01.02 develops and defines corporate and operations strategies. When a firm previously has gone through this process, this task simply may be clarification. When the firm has not previously considered these issues, the task can be arduous.

Chapter 3 included some discussion of operations strategy. The strategy statement required for site acquisition is similar to the example in figure 3.17. It addresses many of the same issues but in broader terms.

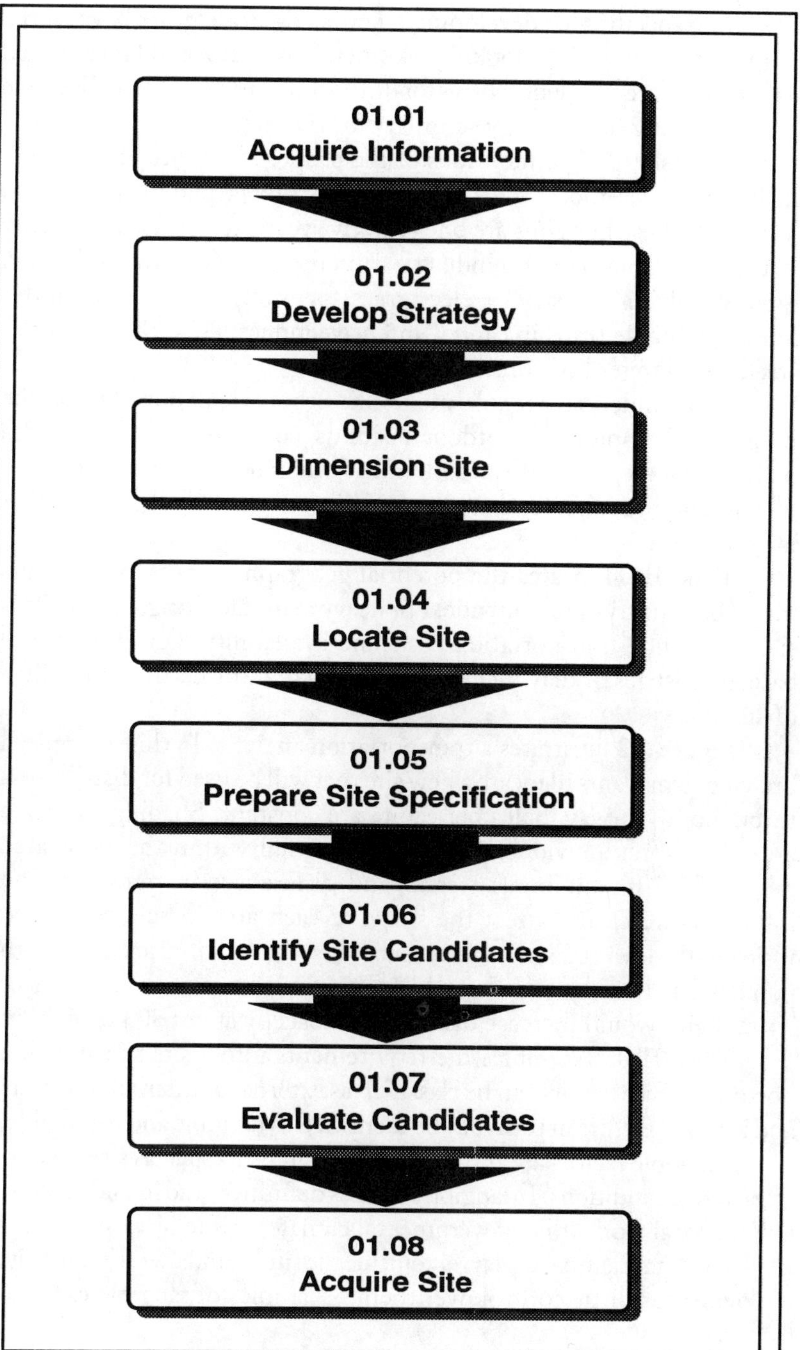

Figure 6.1 - Site Location

The specifics for developing a new site's strategy are beyond the scope of this book. The books by Skinner, Voss, Hill, and Hayes noted as references are excellent sources for further information. The deliverable for Task 01.02 is a strategy summary for the site.

In Task 01.03, dimensions of the proposed site are determined, including under-roof space requirements for several planning horizons. Typically, these horizons are one year, five years, ten years, and twenty years. This varies between industries and firms. Fast-growing firms in rapidly evolving industries—electronics, for example—work on shorter horizons. Stable firms in capital-intensive industries, such as paper or steel, have longer horizons.

From under-roof requirements and other calculations such as the number of employees, outdoor space is computed. This includes parking, storage, and green areas. Dimensioning also includes a definition of corollary requirements such as utilities, local skills, and other infrastructure.

Task 01.04 locates the potential geographic limits. Such limits might be country or regionwide. For heavy manufacturing, this analysis should include transportation costs and availability. For knowledge-based industries, it may include availability of critical skills or quality-of-life issues.

Figure 6.2 illustrates a transportation analysis. In this example, a brewing firm wants to locate a new site that will be used for distribution throughout Norway. Sales per capita are constant. Shadings indicate areas where nationwide distribution costs fall within the associated ranges. The mountainous terrain and differences in transportation infrastructure clearly affect the shape of each area. The area around Oslo has the lowest distribution cost because of its central location, high population density, and favorable terrain. A location in Bergen or Trondheim would increase distribution costs by about 50 percent.

Task 01.05 assembles the requirements into a site specification document. Site factors can be classified as external or internal. External and internal factors in turn may be identified as tangible and intangible.

Tangible factors are those with physical forms such as site boundaries or soil conditions. Intangibles are less definitive, and include factors like the local work ethic, governmental climate, or site history.

External factors exist largely outside the site boundaries. Frequently, the owner has little control over them. Examples of tangible external factors are:

- transportation proximity;
- site access limitation;
- future street widening;

- airport flight paths;
- neighbor activities;
- runoff/flooding;
- prevailing wind;
- freight transportation;
- residential proximity;
- public transportation; and
- acquisition price and terms.

Intangible external factors might include:

- zoning;
- building codes/permits;
- political climate;
- police and fire protection;
- neighbor attitudes;
- site and area history;
- community power structure;
- manpower;
- natural resource restrictions;
- environmental issues;
- grants and aids;
- tax exemption;
- local taxes;
- political stability;
- economic stability;
- industrial tradition;
- labor laws;
- construction costs; and
- cost of living.

Internal factors exist largely within the site boundary. The owner may have substantial ability to modify or mitigate the effects of them. Some tangible internal factors might be:

- topography;
- soil conditions;
- existing buildings;
- existing traffic areas;
- utility systems; and
- water table.

Some intangible internal factors are:

- landscaping;
- security;
- safety;
- appearance/image;

- managerial style; and
- tract shape.

Focusing on each sub-class reduces the possibility of overlooking an important factor.

The site specification gives real estate agents and site-search experts the information they need to screen potential site candidates. This activity takes place in Task 01.06; the result should be a list of two to eight candidates.

Task 01.07 evaluates the site candidates. Tools for evaluation include weighted factor analysis, financial analysis, simulation, and material flow analysis.

Qualifiers are the absolute minimum requirements for the site. Deciders are features that can vary within reasonable limits. Qualifiers might include items such as: a minimum of 4.5 acres of land or 10,000 KVA electrical service. Deciders might include additional options for expansion land or lower electrical rates. All candidates that pass the initial screening should meet the qualification requirements but will vary with respect to deciders.

To assist with the evaluation, preliminary site plans may have to be prepared. For example, a site that is otherwise desirable may have an odd configuration. A site plan can determine whether the site is suited for the planned buildings and operations. The site's possibilities and limitations must be analyzed. While most evaluation factors are in the original site specification, some may only become known as the sites are investigated.

Acquisition (Task 01.08) includes negotiation and purchase. These negotiations often involve owners, lawyers, environmental specialists, local government officials, and development officials. Union representatives may also participate. The negotiating process may change the decision factors and force a reevaluation. When acquisition is complete, space planning can begin.

Site planning

A site plan shows land use—today and in the years to come. Buildings, open space, parking, and green areas are delineated. The plan reflects the physical results of the business strategy, showing space utilization, transport routes, utilities, maintenance systems, and expansion possibilities.

Space planning for Level 2—the site—positions buildings, parking lots, drives, utilities, and many other site features. Industrial engineers often perform this task jointly with an architect, civil engineer or facility engineer. This task is best accomplished after a specific site is acquired.

However, having a preliminary site plan may help during the acquisition process. The task also may involve replanning an older, existing site.

A good site plan assists with orderly expansion and growth. It helps place short-term layout and facility decisions into a long-term strategic perspective. It is essential to the long-term mission of a facility.

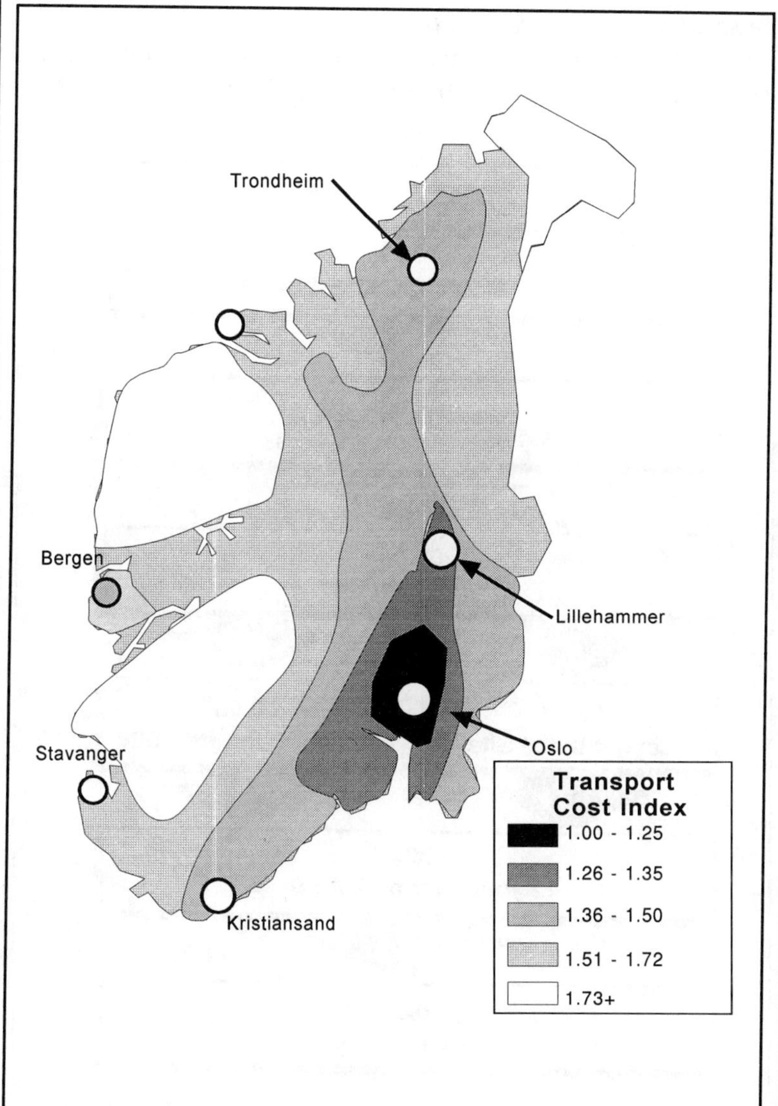

Figure 6.2 - Transportation Analysis

The fundamental macro-space-plan elements, SPUs, space, affinities, and constraints, apply at the site level. However, there are differences between site-level planning and space planning at the macro-level. Site-level space usually means exterior dimensions and exterior space use. Affinities are simpler, and constraints dominate the layout of most sites.

The site-level model project plan

Figure 6.3 is the model project plan for a single-building site. As mentioned previously, large, complex, multi-building sites use a different

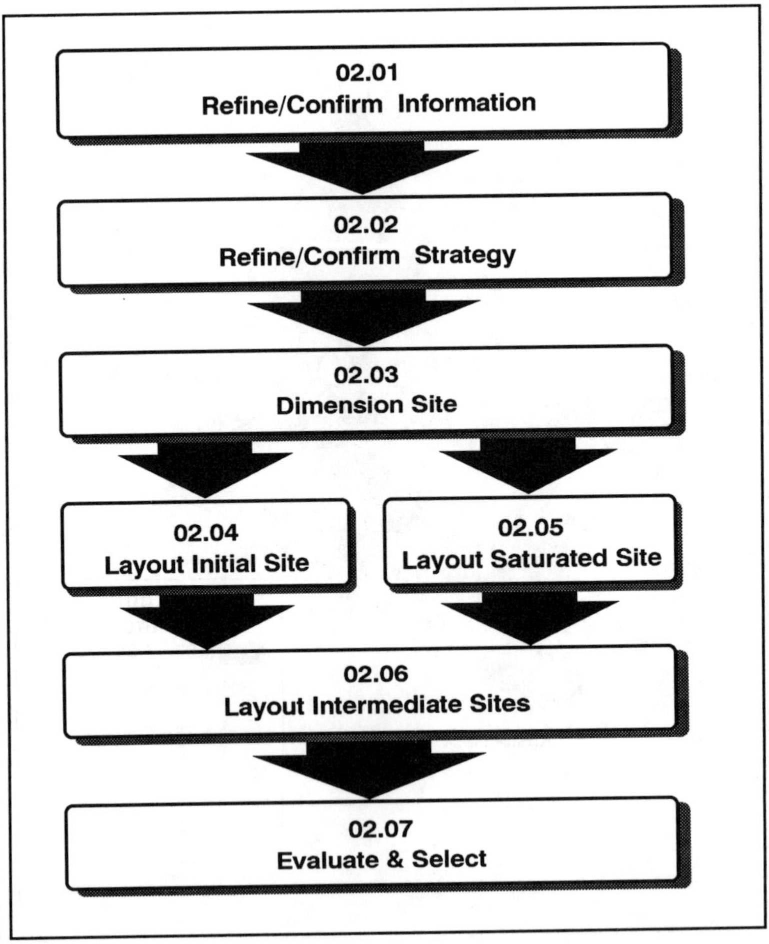

Figure 6.3 - Model Project Plan

approach and considerable site planning experience is needed. Such complex sites are beyond the scope of this discussion. For additional information, see Wrennall and Lee, *Handbook of Commercial and Industrial Facility Management* or Muther, *Systematic Planning Of Industrial Facilities.*

Information needed for the site plan is obtained during Task 02.01. Much of it may already exist if the planners are progressing from the site acquisition phase.

Task 02.02 of the model project plan develops operations strategy. This may simply be a confirmation of earlier work, possibly with additional detail more suited to local conditions. The deliverable for Task 02.02 is a strategy summary similar to the one illustrated in figure 3.18.

Once the background information has been collected and the operations strategy is in place, the site is dimensioned (Task 02.03). This process defines, analyzes, and quantifies those factors that affect key features at each planning horizon, including:

- space requirements by class;
- number of employees;
- shipping volume;
- production volume and mix;
- utility requirements;
- organization structure;
- traffic access; and
- design constraints.

The Initial Site and Site Saturation Plans

The initial site is planned in Task 02.04. For a new site, this would be the first building phase. For an existing site, it is the current configuration. Figure 6.4 illustrates for Vinyl-X, an extruder of vinyl products.

Task 02.05 is the site saturation plan, which places the maximum production space on the site. Also included are appropriate support spaces such as docks, parking, offices, and warehouses. Figure 6.5 shows the site saturation plan.

After an initial plan and a site saturation plan are in place, intermediate development plans can be filled in during Task 02.06. Figure 6.6, when joined with figures 6.4 and 6.5, completes the development plan for the Vinyl-X site. A written plan should accompany the graphic layouts developed during Tasks 02.04 through 02.06.

The model project plan shows Tasks 02.04, 02.05, and 02.06 as individual tasks. However, designers probably will work on layouts for all three at the same time, creating several distinct sets of site plans. In Task 02.07, space planners evaluate these options and select from

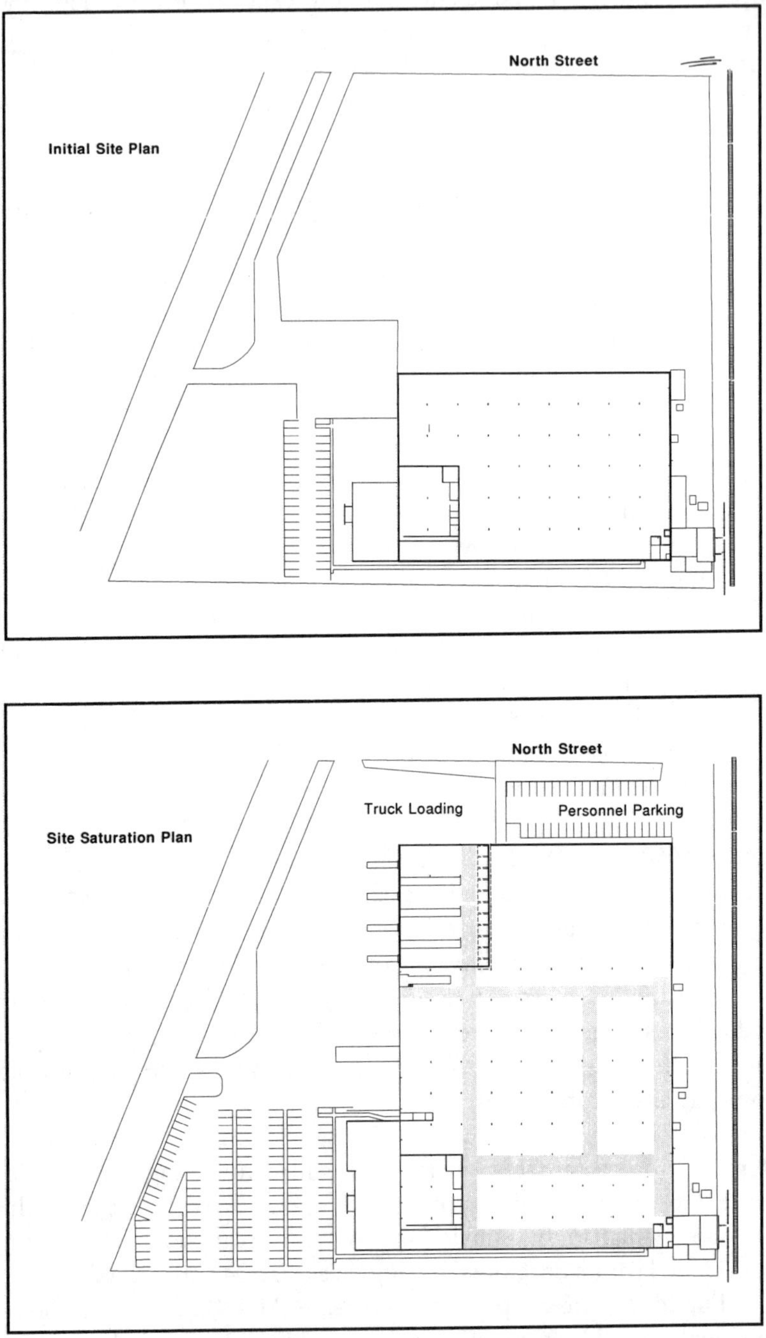

Figure 6.4, 6.5 - Site Plan Development

among them. Decision tree analysis is a viable addition to the usual panoply of evaluation tools.

Other Site Planning Issues

While developing the initial site saturation and intermediate plans, space for immediate use is laid out first. Future space plans also must be anticipated and the transition costs minimized. For example, loading docks are expensive to build. Therefore, they should be placed where they will not be affected during future expansions. Conversely, light assembly workstations are easy to move. Their position is not as critical.

Space flexibility, or the cost of moving a space and the complexity of the facility, must also be considered. Figure 6.7 illustrates. The SPU's flexibility is low if it has special facilities and represents very heavy production. The SPU's flexibility increases with lighter production and simpler facilities.

For the development plan, designers can choose from several types or variations: *zone*, *block*, *duplicate*, and *stripe*. Figure 6.8 shows these space plan types.

A zone plan begins with a central core of space. Each SPU, function, or activity has a related zone. As space needs increase, the SPUs expand outward but within their designated zones. Zone plans

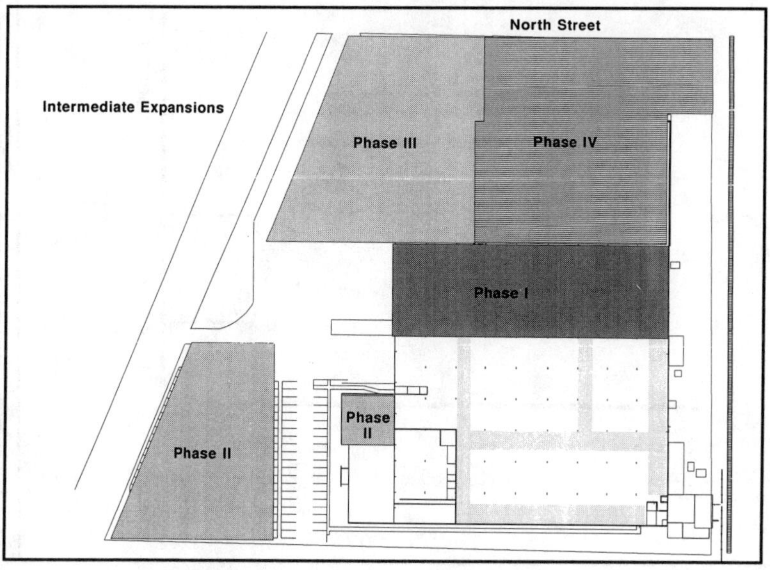

Figure 6.7 - Site Plan Development

allow orderly growth without altering production organization. However, as SPUs become larger and larger, coordination and control difficulties may develop, especially if growth also brings more product variety.

A block plan begins with blocks of land designated for each function or area with long travel distances between SPUs. Each building may then expand within its block. Block plans work well when each SPU is a largely self-contained entity.

A duplicate plan starts with an initial facility, which is then cloned on adjacent land. This avoids the problem of increasing scale; however, it means capacity increases in large increments.

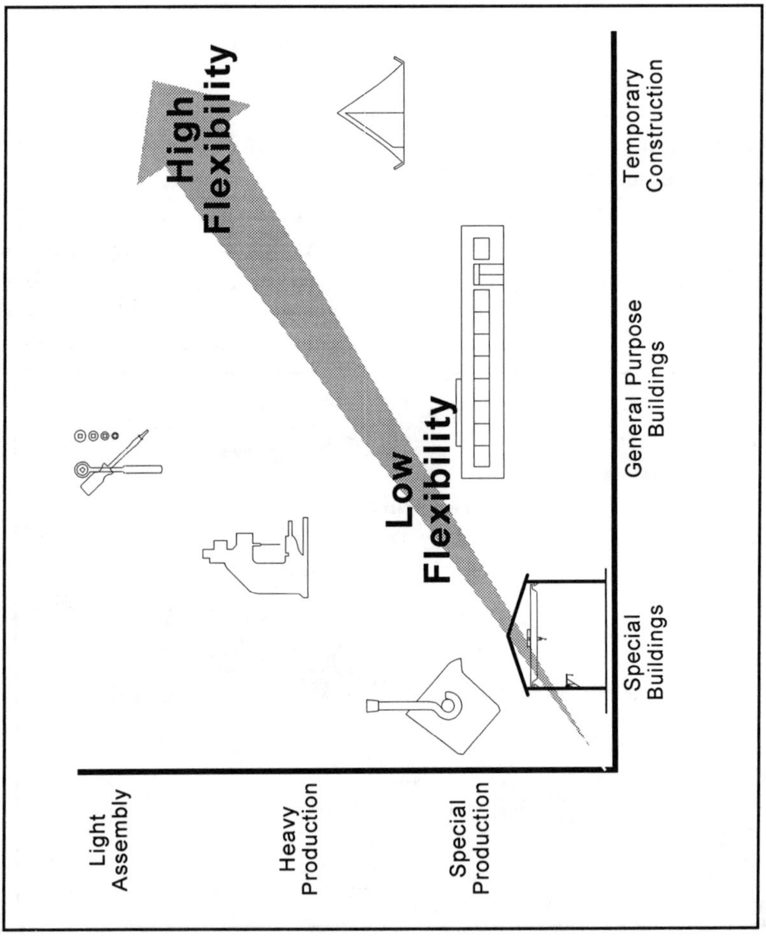

Figure 6.7 - Process and Facility Flexibility

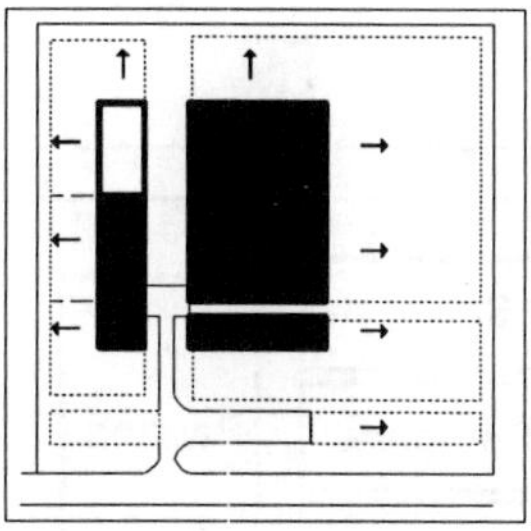

Zone Plan

Areas are reserved for expansion from a center core outwards.

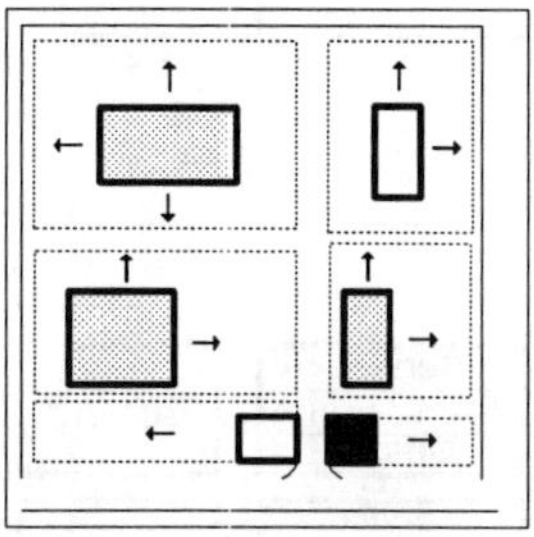

Block Plan

Areas are blocked out by function or space type. Each facility expands within a designated block.

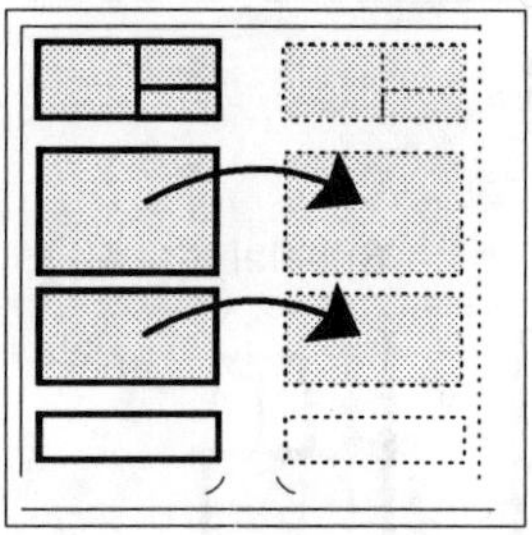

Duplicate Plan

Each facility has balanced space. Expansions duplicate the original.

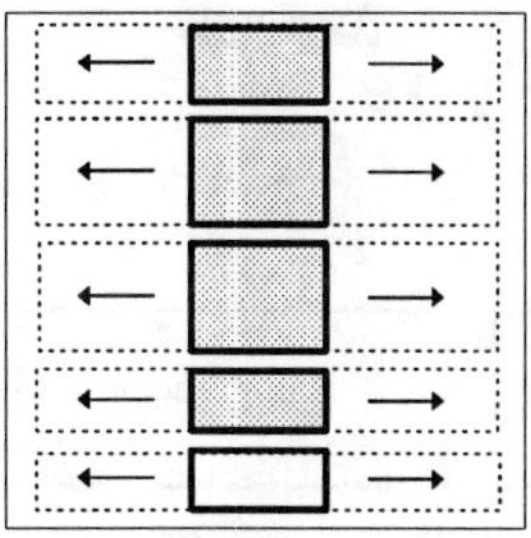

Stripe Plan

Areas are reserved in stripes across the site. This is a special type of zone plan.

Figure 6.8 - Site Development Styles

A stripe plan is a special version of a zone plan where each SPU expands horizontally. This is a good approach for simple, linear processes that expand as new production units are added. It may not work well for more complex products and processes.

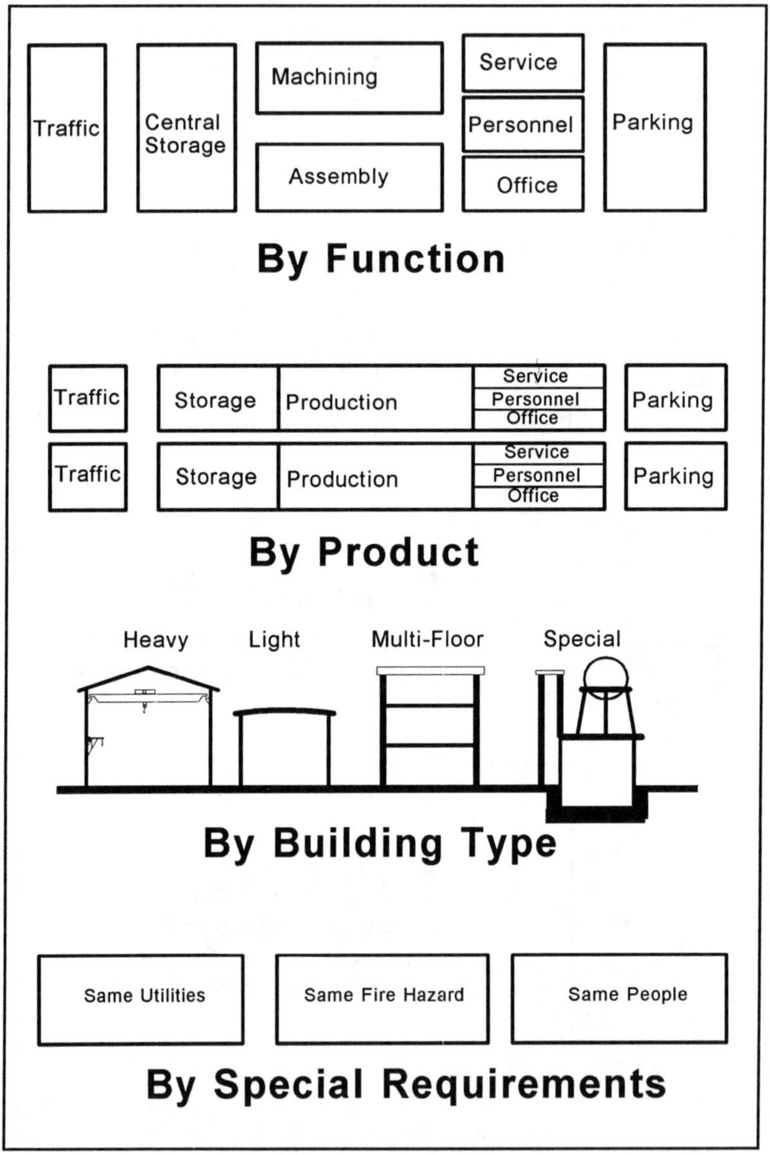

Figure 6.9 - SPU Grouping

The Site Plan

The planning process for the site plan is similar to the process for macro-layouts. The first step is identifying the SPUs. Related activities or features are grouped into logical SPUs based on function, product, type of building, or special requirements(Figure 6.9). Figures 6.10 through 6.14 use BCD Industries, Ltd., as an example. Figure 6.10 lists the names and acronyms of each major SPU as well as the initial space requirements. If an existing site were being reconfigured, current space would be shown at this point.

SPU Summary

	Phase I	Phase II	Floor Load	O-Head Clear(m)	Special Installation	Expansion	Flexibility	Building Type
Cutting	2300	2000	Normal	12	Crane	Small	Moderate	Heavy
Light Machining	2900	4000	Normal	6	Crane	Medium	Moderate	Normal
Heavy Machining	1700	2300	High	12	Crane	Medium	Low	Heavy
Assembly	3500	5200	Normal	6		High	High	Normal
Raw Matl Storage	2600	3500	High	12	Crane	Medium	High	Heavy
Parts Storage	1100	1700	Normal	6		Small	High	Normal
Finished Storage	4600	6900	Normal	6		High	High	Normal
Technical Service	400	600	Varies	Varies	Utilities	Very Small	Low	Special
Office	1200	2400	Light	3		Medium	Moderate	Multi-Flr
Personnel Facilities	800	1700	Light	3		Medium	Moderate	Multi-Flr
	21100	30300						

Figure 6.10 - SPU Summary

Figure 6.11 is the affinity chart that identifies the affinities between the SPUs in the site plan. Some of the SPUs, such as rail siding and road, are site features. This is typical for site level planning.

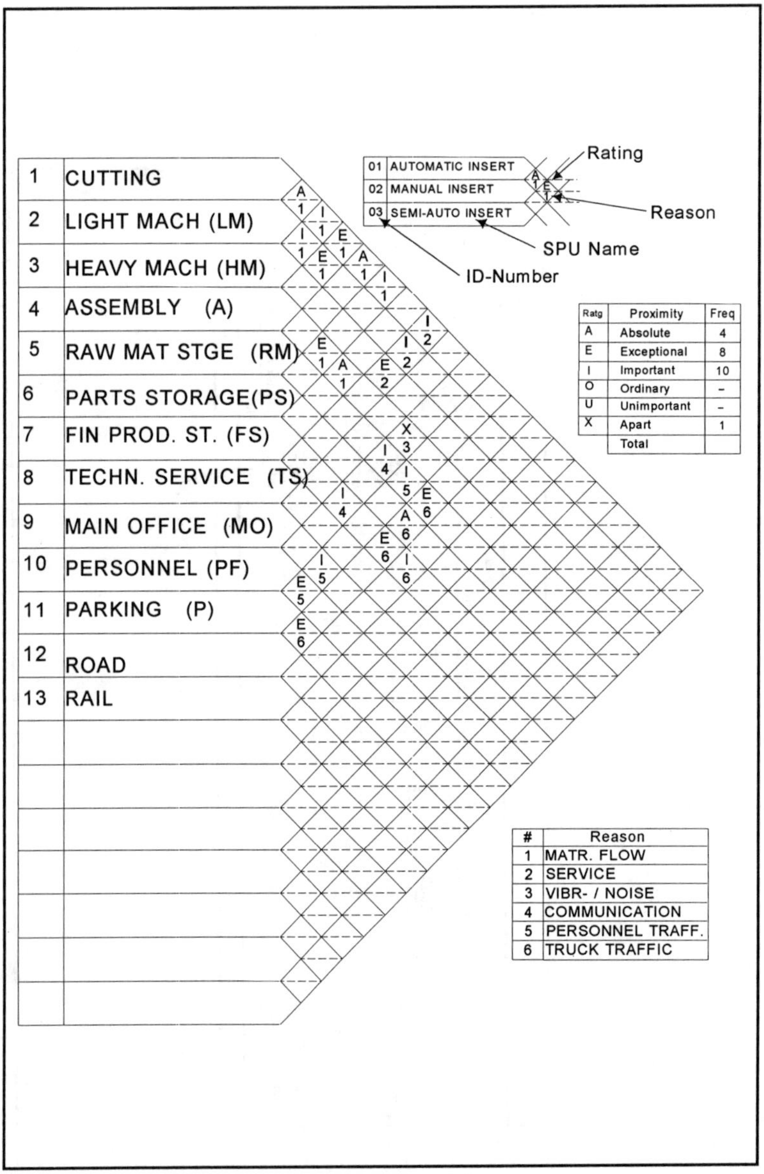

Figure 6.11 - Affinity Chart

Next, a configuration diagram and the layout primitive should be constructed (fig. 6.12). Figure 6.13 shows the same diagram coded for flexibility and building type. Coded diagrams help in the preparation of the site saturation and development plans.

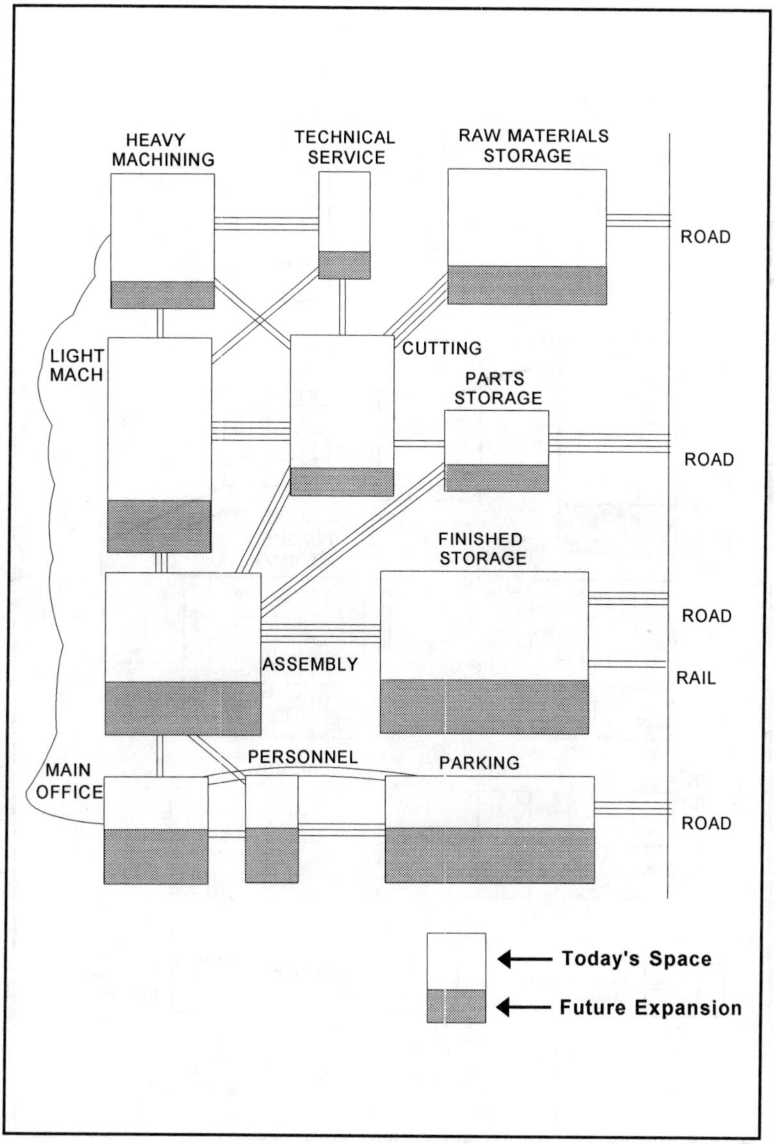

Figure 6.12 - Layout Primitive With Expansion

BCD's space plan is shown in figure 6.14. It consists of an initial space plan with a subsequent expansion. The top portion of the illustration shows operating departments at stage I and stage II. The

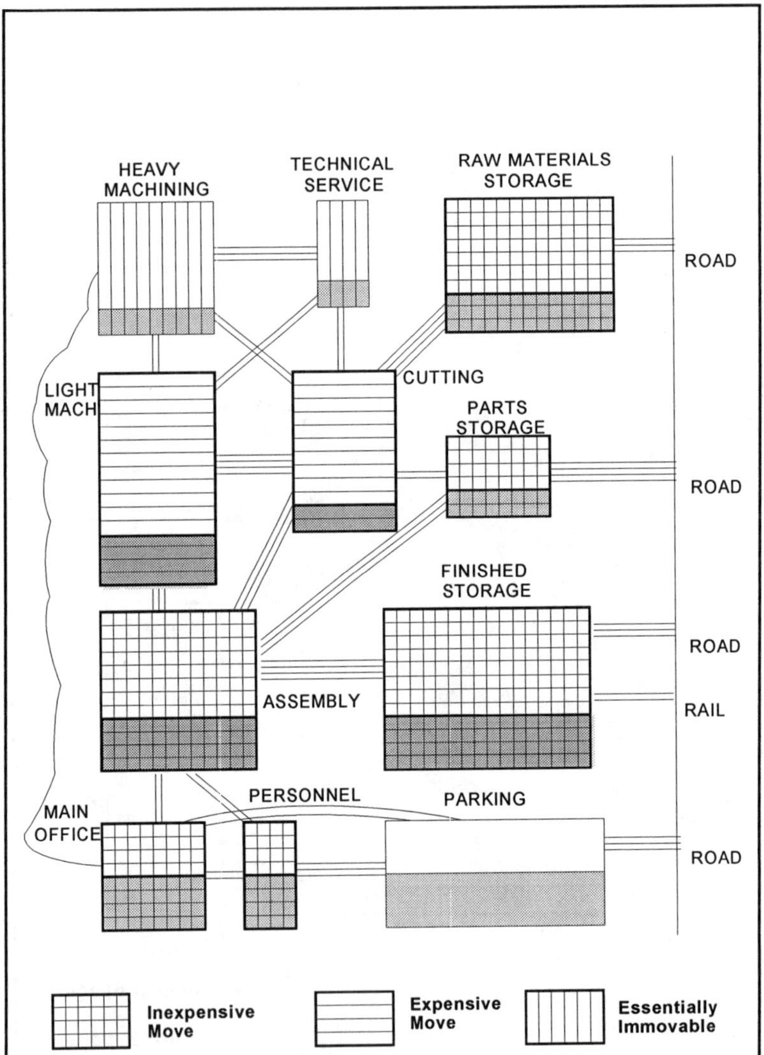

Figure 6.13A - Layout Primitive Coded For Mobility

middle portion shows the type of building. The bottom portion shows the mobility of space at both stages. In the expansion, few SPUs require movement. None of the less mobile areas require movement.

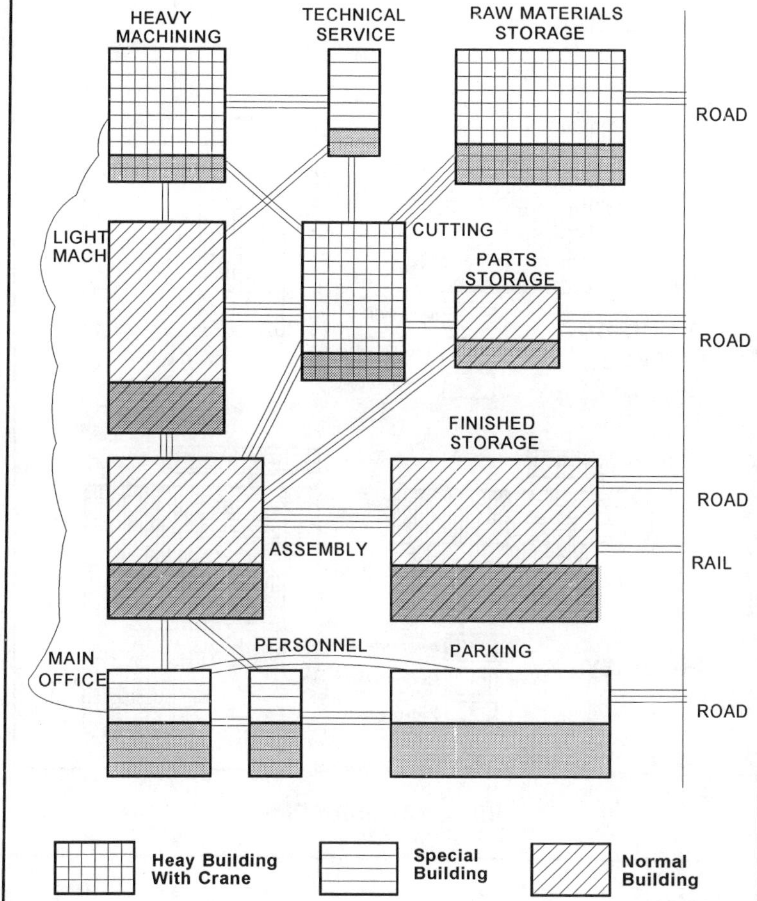

Figure 6.13B - Layout Primitive Coded For Building Type

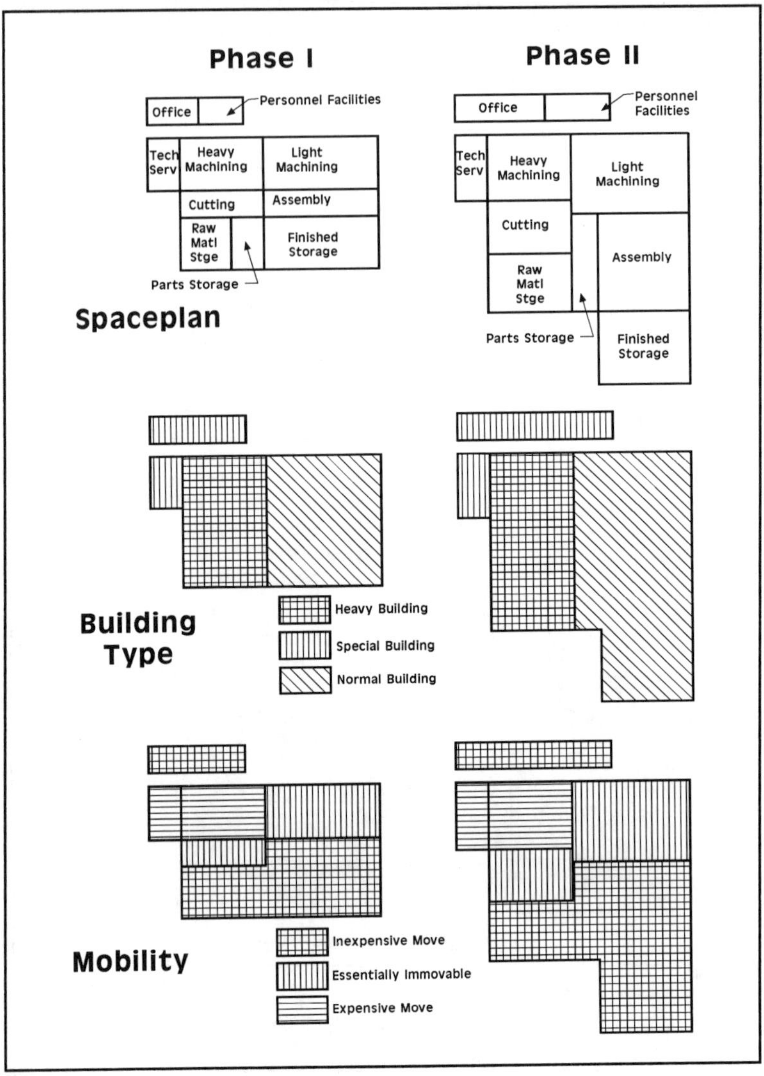

Figure 6.14- Space Plan

Conclusion

This chapter has set out the fundamentals of site selection and site planning, demonstrating how these detail levels fit into the total facility planning cycle. Model project plans outline the steps required to select and plan a small, single-building site. When all the tasks are completed, the result is a plan that should please those individuals who participated in the process.

Chapter 7

Office Space Planning

Written by Herbert Tuttle

The shift from manual labor to knowledge-based work has increased the need for productive offices. Office personnel are no longer simply typists and clerks. Office work now also includes accounting, administration, manufacturing support, information services, human resources, engineering, customer services, and team meetings. It includes many types of knowledge-based and information-intensive work. Simultaneously, manufacturing operations are becoming cleaner, smaller, and more sophisticated. Many factories are cleaner and more user-friendly than typical offices. The people who make products often work elbow to elbow with engineers, schedulers, and designers who now sit on the manufacturing floor.

The distinction between product and service has blurred. Service providers use the manufacturing principles of Henry Ford. Manufacturers include a package of services as part of their products. These may include special designs, customization, financing, repairs, or application assistance.

Flexibility in manufacturing has spread to flexibility in office layout and design. It is not uncommon to rearrange entire floors of offices in a weekend. Many advances in furniture, telephones, and computer system designs have been made. These new systems provide design flexibility, are easy to install, and can accommodate the increasingly rapid changes in the workplace.

Office structure has undergone considerable change. The office may now have teams rather than functional departments. These temporary or permanent teams may manage a process or project. The office structure and company culture affect office planning and design.

The office being designed will serve people with various needs. Their performance—be it clerical, administrative, or intellectual—depends in part on how they feel in their surroundings. Designers usually want the employees to be involved in the design of the layout, as well as the selection of furniture and fixtures. When users help determine their environment, they feel more comfortable in it, which in turn contributes to their motivation.

Every employee may not agree with every aspect of a particular layout. It is possible, however, to get a group consensus. People usually can agree to adopt the chosen layout.

This chapter gives a designer or design team the foundation for creating an office space plan. The fundamentals of office space planning remain the same whatever the size or location of the offices. Included are approaches, layout alternatives, and hazards. It demonstrates how a team may be used to develop an office layout acceptable to all users.

Approaches to office planning

Chapter 1 discussed common approaches to space planning: experiential, master building, cloning, bottom-up, systematic, and strategic. Office planning frequently uses the same approaches, along with others such as interior design, organization-based, space rationing, and information flow.

The *interior design* approach focuses on artistic design and creativity. The outcome is a pleasing environment, but it may not be the most efficient. It has merit for executive offices and reception, lobby, and meeting areas. Typically, it is expensive and rarely contributes to productivity.

The *organization-based approach* is the most common. It superimposes the organization chart over the floor plan, starting with the president or top executive in a corner office with disproportionate space. A secretary sits nearby, and those managers reporting directly to the executive are along outer walls. Support staff is then shoehorned in outside the executive offices as illustrated in figure 7.1.

The *space rationing approach* is often part of organization-based space plans. Rank in an organization determines work space size. Each person has a specified area corresponding to their position in the corporate hierarchy. They often have specific equipment and furniture. Corporate leaders, often in remote areas, usually determine the space standards.

Layouts based on hierarchy may allocate windows effectively but rarely contribute to productivity. They facilitate communication, but only along official channels. Hierarchical layouts rarely fit in a team environment.

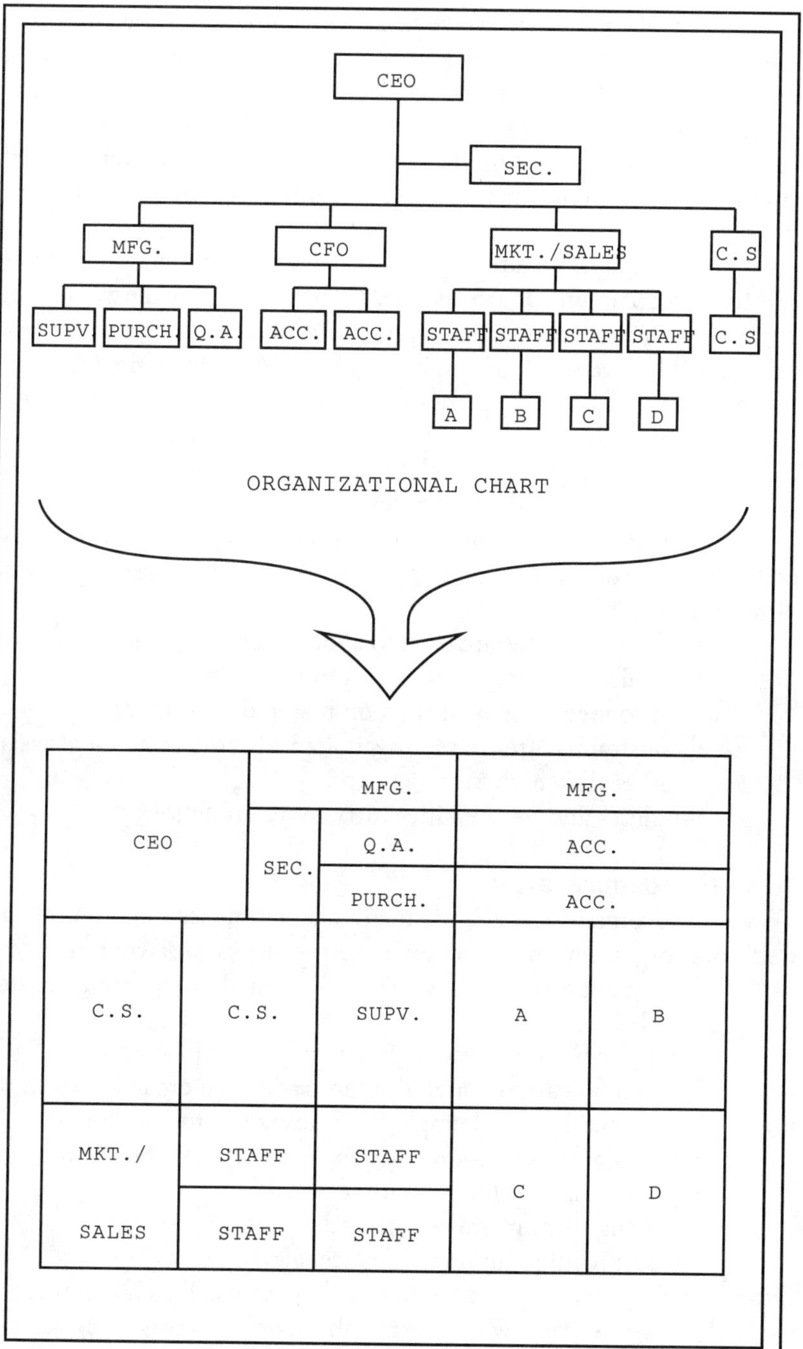

Figure 7.1 - The Organization-based approach to Office Planning

Many office layouts evolve from the *information flow* approach. Two paradoxical trends are influencing this approach. First, communication technology is reducing the importance of physical location for some types of information. Second, team-based organizations are increasing the importance of oral and interpersonal communication. This interpersonal communication is sensitive to physical location.

Space planning for an office uses the same principles and procedures as other space planning. It incorporates the best features of the traditional approaches, using the same principles and procedures as macro-space-planning. The levels of detail apply, beginning with global site location. Site plans and macro-space plans follow. The detailed levels of work cell design and workstation design are as important for an office as for a factory. At each level, space planning units, affinities, space, and constraints assist in the design process.

The special needs of office environments change the emphasis and application of the space plan approach. Among the most important differences are:

- space standards are used to allocate and calculate space;
- affinities depend more on communication and personnel:movement than on material movement
- constraints often have a psychological or organizational basis rather than a physical one; and
- politics and personalities carry more influence.

Team-based office layouts

Team-based organizations have many characteristics uncommon to traditional organizations. Characteristics that affect the layout include:

- frequent meetings (twice daily) of small groups for an hour or less;
- the generation of more charts, graphs, or project communication materials than traditional organizations;
- substantial annual employee training (40 to 120 hours);
- frequent presentations to management and other teams;
- project-oriented temporary teams;
- varying team membership; and
- direct team input on workplace layout.

Offices used by teams should be designed by teams. This is congruent with the empowered work team philosophy. However, some organizations have not adopted empowered work teams. An office space planner should realize what is appropriate for the organization.

The office layout team

Team-designed layouts help overcome difficulties that arise from complexity, human factors, and organizational maneuvering. If empowered teams are not used, people from each department should participate in other ways.

When a team does the work, Task 03.01 may require more than project planning. Organizing an office layout team involves expanding Task 03.01 to include the following steps:

1. Appoint a facilitator.

2. Select team members with good interpersonal skills. Preferably, they should have training or experience in teamwork. The membership should represent all major departments and areas.

3. Develop an agreeable team name.

4. Develop a brief mission statement that satisfies all team members. It should state who, what, when, where, why, and how. Here is an example: "The Building Blocks Team for Foxmann Manufacturing Co. will design one office layout by June. It will have the approval of all internal customers and achieve budget goals."

5. Assign responsibility. The team leader assigns tasks and responsibilities. He or she is often an engineer and should coordinate assignments for team members. For example, each person should be the liaison for his or her department or home team. Someone should work with those departments or teams not represented on the layout team. Other assignments include meeting agendas, minutes, contact with suppliers, drawing coordination, sketches, and layout. Someone should be the communicator and report to management and other teams.

6. Agree on team meeting dates, times, and duration.

7. Develop ground rules for the team's operation.

The layout project

For the office macro-layout, a model project plan (see fig. 3.1) can be used. Chapter 4 explains how to design work cells and departmental space plans. Office planners should use the task procedures outlined in both chapters, applying them directly to office layouts. Several tasks probably should be modified, however, so that they meet the requirements of an office project.

Office user survey

Part of acquiring information is a survey of the office space users—the internal customers. This should be called Task 03.09A, "Office User Survey," and should be added to the model project plan.

Figure 7.2 is an example of an office user survey that can be tailored to meet specific needs. Management should review any survey before it is distributed to everyone who will use the office, perhaps even outside customers, vendors, and visitors. To supplement the survey, team members may wish to interview senior executives and other key people.

The survey assists in many subsequent tasks. It can help with space calculations, and it allows planners to identify opportunities for process improvement. It also helps identify constraints. The data collected should be documented in spreadsheet format. The data from the office user survey also should be used to develop affinities and space requirements. These affinities then should be reviewed with many or all of the users so that they may see the rational process behind their layout. This review is an important consensus builder.

Material and information flow

In most offices, material flow is inconsequential; therefore, Tasks 03.12 through 03.14 may be unnecessary in the office planning process. Also, in certain service environments it is difficult to visualize the product. This may mean more diligence, focus, and effort than it would where the product is tangible.

Information flow, on the other hand, may warrant significant analysis. Acquisition of the latest communications technology may eliminate a company's traditional needs for proximity. The purchase of a new computer system, for example, could significantly alter layout requirements.

It may be worthwhile to develop a task that deals with key information processing issues. The task could be substituted for Task 03.20, "Handling Issues."

Constraints

There are specific office planning considerations and characteristics that affect the office planning process. These are: multi-use areas, communications systems, shared services, safety and security, furniture, interaction, working hours, decor, lighting, greenery, new construction,

To: Ms. Deb Jones
From: Building Blocks Team, Foxmann Div. of Arron Elf, Inc.
RE: Office Design (layout or redesign) Survey

We need your help in designing an efficient new office layout. Please complete this survey and submit to Jane Plow of your home department.Jane Plow is the representative of your area on the Office Design Team. We need your survey by February. Please add any additional comments.

1. What are the three departments (teams, areas) with which you have the most interaction?
A ________________ B ________________ C ________________

2. In the Deparmtnets listed in Number 1 above, what are the names of the people or titles of the people with whom you have the most frequent contact?
A1 ________________ A2 ________________ A3 ________________
B1 ________________ B2 ________________ B3 ________________
C1 ________________ C2 ________________ C3 ________________

3. Please check the items you use daily.

__ Desk	__ Telephone	__ File	__ Adding Machine
__ Work Surface	__ Computer	__ Speakerphone	__ In Basket
__ Bookcase	__ Fax	__ Copy Machine	__ Conference Room
__ Desk Light	__ Others (List)		

4. Please check the items you use, but not every day.

__ Desk	__ Telephone	__ File	__ Adding Machine
__ Work Surfaces	__ Fax	__ Copy Machine	__ Conference Room
__ Computer	__ Speakerphone	__ In Basket	__ Bookcase
__ Credenza	__ Desk Light	__ Others (List)	

5. Check what you have now.

__ Office with locking door	__ Office with non-locking door
__ Office with no door	__ Cubicle, tall walls (>5 feet)
__ Cubicle, medium walls (<5 feet)	__ Open or low-walled cubicle

6. What percentage of the work week do you spend in meetings in your work space (office)? __________

7. What percentage of the work week do you spend in meetings outside your work space (office/cubicle)? __________

8. What percentage of the work week do you spend in meetings away from your work space (office/cubicle)? __________

9. If your work space were to be changed, what would you like to see arranged differently?
____
____

10. What do you dislike about the present office layout or arrangement?
____
____

11. What do you like about the present office layout or arrangement?
____
____

12. What is the most important aspect of office layout the Office Design Team should consider?
____
____

Figure 7.2 - Example of an Office User Survey

rest room and utility areas, and reception areas.

Multi-purpose areas are important for any office. The utility of these areas is often over-looked. They are the first areas to be eliminated due to budgetary constraints. Multi-use areas serve as private meeting rooms, conference rooms, team meeting rooms, training facilities, and temporary work areas. They need writing surfaces such as dry-erase boards or chalk boards. Separate light control switches, separate telephone jacks, and data jacks should be installed. Sliding doors should be built in as panels. Small tables are desirable additions as well. They may be used as training desks or work tables; several may be pulled together to create a conference table. The equipment for multimedia presentations also should be accommodated. See the example of a multi-use area in figure 7.3.

Plans for information systems and telecommunication are necessary in the office layout design. Most people will need telephones, computers, or other devices. Wires and connector devices also should be considered. A personal computer or telecommunications installer should be invited to a team meeting for input. Shared devices like faxes and small copy machines should be located in a central location.

Planners should develop an office security and safety plan. Video monitors, metal detectors, and other devices can enhance security at low cost. "Secured" areas such as laboratories, archives, and computer equipment should be part of the plan. Power failure, fire, earthquake, rising water, tornadoes, civil disturbance, bomb threat, strikes, and other emergencies should be considered.

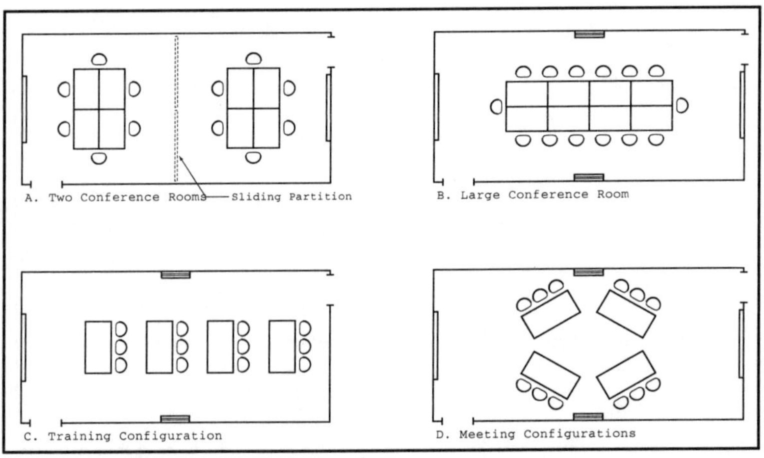

Figure 7.3 - An Example of a Multi-use Area

Office space planners need to develop a relationship with a supplier of furniture and office fixtures that has provided competitive bids and quality service in the past. Ideally, office situation representatives should be made part of the office planning team. Figure 7.4 is a check list for office layout.

In a multi-floor office situation, people who interact with groups on other floors should be located near a stair or elevator. Cubicles, work areas, and traffic areas should be integrated. Making the office comfortable as well as productive is the goal. Designers must allow for flow in horizontal and vertical dimensions.

Flexible hours and job sharing are becoming more common in the work place. Typically, flexible hours mean employees must be at work for a set of core hours (for example, 10:00 a.m. to 2:00 p.m.) or they may work any pattern as long as they work at least eight hours a day or forty hours a week. With job sharing, two or more people share one job. This work force flexibility, while seemingly unrelated to office planning, is a worthwhile consideration.

As more companies adopt flex-time, job sharing, casual overtime, and contracted projects, the special circumstances they need and cause should be considered when planning an office. They affect peak facilities usage and on-the-job safety and security. They should be considered when scheduling janitorial or maintenance services. It is difficult to work with the roar of vacuum cleaners outside a cubicle.

Most employees want natural lighting and windows that open for fresh air. These are not always possible. Economical heating and cooling may outweigh the need for individual control of windows. Window blinds or shades should be provided. Again comfort and productivity are complementary here.

Lighting position and intensity should correspond with the configuration and position of desks and computer terminals. Various covers or diffraction devices can control lighting in computer areas. In some cases, task lighting works well.

Plants add much to an office and should be part of the layout design. Large plants can provide geographic barriers and add fragrance and color. It is an eye for such details that makes the office attractive yet functional.

The design of a new office is an opportunity to correct problems in an existing office and to accommodate future growth. However, an office layout that allows for new construction requires long-range planning. Staffing levels over the next ten years are top priority. Without a growth plan, offices become crowded and inefficient. Plans for expansion without disruption to the office also should be made.

Rest rooms, closets, and utility areas are a necessity for every office. The building architects determine such locations, usually before a layout is complete. If at all possible, the office planning team should work with the architects to ensure that these areas are conveniently located for the users.

The lobby or reception area of an office is where the public gets its first impression of a company. An architect or interior designer should work with the space planner(s) on the design, and senior management should be consulted about the image they want to portray. It is a good place to display products or representations of a company's history and its commitment to quality and excellence.

Moving Day

The team has successfully completed a survey, prepared alternative layouts, and received approval for an office layout and budget. But the day of the move or relocation requires planning, too. What files will move, when, and where? What utilities, data, telephone wires will be moved and when? The move should probably take place during a weekend, or holiday, if possible. A "dry run" before the move is a good idea. Upon completion of the trial move, the planning team should discuss it with all involved parties and identify any forgotten details.

OFFICE FURNITURE
— Is there a desk or work space for each person on the organization chart?
— Do the type, style, and color match?
— Do utility runs match in cubicle walls?
ADA REQUIREMENTS
— Can wheelchair-empowered employees, customers, and vendors travel about your office layout?
— How much strength does it take to open the doors into the office?
— Are alarms configured with flashing strobes for hearing-impaired employees?
— How will you communicate the new layout with visually impaired employees?
— Do the restrooms comply with the code and ADA requirements?
MAINFRAME COMPUTER FACILITIES
— Are the floor and ceiling designed to accommodate the additional wires?
— Is temperature control adequate?
— What type of fire sprinkler is available?
RECEPTION OR LOBBY AREA
— Is there security?
— Does the area present the proper image?
SOUND CONTAINMENT
— Who is next to the lunch room?
— Where are the printer and fax machines located?
— Who is next to the conference and training rooms?
ERGONOMIC DESIGN
— Is the workstation designed for a person to spend at least eight hours there?

Figure 7.2 - Example of an Office User Survey

— What are the chair, desk, and keyboard height relationships?
— Is the height of the computer monitor in each work area adjustable?

LIGHTING

— Has the ceiling lighting been relocated to correspond to desks, tasks, and aisles?
— Are the levels adequate for meeting and training rooms?

HEATING, COOLING, AND CLEAN AIR

— Will a change in load reduce efficiency?
— Will the air flow pattern need to be changed?
— Did you clean the ducts?
— Where are the thermostats?

STATIC ELECTRICITY

— Where is the floor covering?
— Are the work surface and cubicles grounded?

MAIL ROOM

— Is it secured?
— Who has access?

LUNCH ROOM

— Are there tables and chairs for all?
— How and when will the vending machines be serviced?

RESTROOMS

— Do the number and type of fixtures meet zoning codes?
— Can they be easily maintained?
— Where are the supplies stored?

FLOOR LOADING

— Is the cumulative weight within limits?
— Is the sub-floor for the mainframe computer adequate?

EXERCISE AREAS

— Is the ventilation adequate?
— Is there adequate locker and rest space?

TELEPHONE EQUIPMENT

— Is the ventilation adequate?
— Is the electrical service adequate?
— Is it really large enough?

EMPLOYEE ENTRANCE

— Is there a time clock or sign-in area?
— What is the security?
— Are there cameras?
— Is there a waiting area?

HUMAN RESOURCE OFFICE

— In an adjacent plant, is the Human Resource Office nearest to the plant?
— Is there privacy and confidentiality?

MEDICAL FACILITIES

— Are they located in the area most likely to use them?
— Are they accessible by a wheelchair and stretcher?

JANITORIAL FACILITIES AND SERVICES

— Is there adequate storage space?
— Is it adequately ventilated?
— When will it be used?
— When will cleaning be performed?

LIBRARY

— What is the lighting level?
— What is the noise level?
— Is there space available for ten years of periodicals?

CLOSETS AND STORAGE

— What will they be used for?
— Will they be adequate in ten years?

Summary

This chapter has provided the basic elements for organizing a office layout team, conducting a survey of internal customers. The relationship between facilities planning and office space planning is illuminating. Using a macro-level approach to office space planning makes designing office layouts easier. Checklists and examples assist with an activity that may only occur once in a professional career. These techniques and tools are designed to help create layouts that are logical, flexible, and meet the needs of internal customers.

Chapter 8

A Final Note

In this book we have tried to cover a complex and ambiguous subject. We suggest approaches and methods that allow relatively inexperienced industrial engineers to design good facility plans with efficiency and effectiveness. The methods of this book will also help experienced facility planners.

Chapter 1 placed facilities and the planning process in historical and business perspectives. It explained why facilities are so important. It traced some general trends in facility design. It looked to the future of working facilities.

Chapter 2 established a framework for the planning process. It defined levels of detail that correspond to planning phases. It explained the four fundamentals of space planning: units, affinities, space and constraints. It defined conventions useful in all phases of planning.

Chapter 3 demonstrated how to design a macro-layout. Macro layouts size and locate departments, areas or other large-scale features within a building. This is the layout phase that is usually most important for industrial engineers. We introduced the Model Project Plan to guide designers through their tasks. We examined the important essentials of an operations strategy. We explained how to synthesize information and strategy into workable facility layouts.

Chapter 4 addressed the details of equipment and furniture location. It presented a methodology and principles for designing effective work cells.

In Chapter 5, we examined the lowest detail of space planning—the workstation. Here we explained how to design the workspace for efficiency, effectiveness, safety and health.

Site planning was addressed in Chapter 6. The special issues of office space planning were explored in Chapter 7.

We would like to leave you with some key points to remember as you work through your next facility project:

- isolate the levels and focus on only one level at a time;
- try to work from the highest to the lowest level unless special conditions dictate otherwise;
- remember the four fundamentals: SPUs, affinities, apace and constraints;
- follow the Model Project Plans. Accomplish each task in sequence and do not look too far ahead. Uncertainties will eventually be clarified;
- know thy process;
- remember that people are an integral part of every work process;
- the most successful space plans are interactive and participative projects. Users supply information, insight and needs; management develops strategy; the designers synthesize a range of workable plans; and all participate in the evaluation and selection; and
- a clear and appropriate operations strategy is your beacon in the night.

Glossary

activity areas. (See space planning units.)

affinities. Various factors that demand closeness between any two cells in a space plan. For example, communication or personal interaction. In workstation design, affinities evolve from the movement of parts, movement of hands, and the necessity for observation, and other indirect factors.

affinity diagram. An idealized spatial arrangement that eventually becomes a space plan. Also called a configuration diagram.

affinity zones. Locations in a workstation that are within an operator's optimum visual and reach zones.

allocation of functions. Divides work between people and machines and largely determines the quality of the operator's work experience.

anthropometry. The study of the dimensions, weights, and strengths of human body segments.

biomechanics. The study of mechanical forces in human movement, including the interaction between individuals and their physical environment.

block plan. In site planning, a development plan that begins with blocks of land designated for each function or area with long travel distances between SPUs. Each building may then expand within its block.

bottom-up. Facilities planning that starts with the details. Departmental units are designed first and, eventually, the overall facility plan is built.

broadcast. In this system for coordinating production, a schedule dictates the rate and product mix for the final operation.

business architecture. The design and development of facilities, organization, products, and processes into a functioning, interactive

business system. Also known as manufacturing strategy and corporate reengineering.

cells. (See space planning units.)

channel route structure. A fixed route with various stops along which equipment moves. At each stop loads originating at that point are picked up and those destined for that point are dropped off.

circulation. A material flow method that uses mobile operators to carry a product or transfer a batch through separate operations.

cloning. Duplicating an existing facility or portion of it.

composite cell. A work cell made up of several smaller cells.

configuration diagram. (See affinity diagram.)

constraints. Those conditions that limit an ideal space plan. For example, building size and shape, columns, floor loading, utility configurations, and external features.

corporate reeingineering. (See business architecture)

direct link. A physical link between processes that allows products to move from the first to the second process without queues, buffers, or delays. The processes start and stop together.

direct route structure. A route structure that allows material to move from its origin directly to its destination.

duplicate plan. In site planning, a development plan that starts with an initial facility, which is then cloned on adjacent land.

dynamic lot size. The balance between operations from work piece to work piece.

dynamic work. Tasks that require an operator's body to be in motion.

elemental calculation. One of six methods used for calculating space requirements. Each piece of furniture or equipment assigned to an SPU is measured and added together for the total amount of space.

experiential facility planning. Planning facilities based on past experience, common sense, and instinct.

external factors. In site planning, those site factors that exist largely outside the site boundaries.

FacPlan. A facilities planning model that focuses on both strategic issues and minutiae, using a model project plan to guide and structure each project.

floating task assignments. The hierarchical assignment of primary and secondary tasks. As the workload fluctuates, the operator may move from the primary to the secondary.

functional cell. A single process work cell that operates on a wide variety of unrelated products.

global site location. The site location level of space planning where the firm decides where to locate facilities and determines their missions.

group technology cell. A work cell in which a series of operations for several products takes place in a single cell.

human-machine system. In a human-machine system, one or more equipment displays show the operator the internal equipment status. The operator alters control settings to change the machine. Continuing observation allows the determination of the effect of altering controls.

internal factors. In site planning, the factors in the site boundary.

intuitive charting/grouping. Examining products, often with process charts, to glean information and assign products to logical groups.

kanban. Japanese word for card. This production coordinations system uses a small mixed stock from which items are pulled frequently. At the time of each withdrawal, the production center is signaled, sometimes with a card.

key manufacturing task. A task that manufacturing must do well to survive in the market.

layout. (See space plan.)

macro-space planning. The space plan level at which the building, structure, or other sub-unit of the site is planned. Designers define and locate operating departments and determine overall material flow. The most important level of planning, it sets the focus, or basic organization, of the factory.

manufacturing focus. The organization of products and processes.

manufacturing strategy. (See business architecture)

master building. Facilities planning that focuses on construction and buildings.

material flow. The movement of parts, products, and/or services from work cell to work cell and throughout the facility(s).

material flow analysis. Using collected data to calculate material flow between each combination of SPU pairs.

material requirements planning. Using computer databases to schedule and track production, plan capacity, collect costs, and provide many other functions.

micro-space planning. The space plan level at which the location of specific equipment and furniture is determined. The emphasis shifts from gross material flow to personal space and communication.

model project plan. A framework of detail levels that translates into planning phases.

motion economy. Steps and procedures that simplify and improve manual work.

non-flow affinities. Intangible factors that give rise to affinities (q.v.).

physical infrastructure. Supports operations for all or most of a product line but does not contribute directly to the process.

physiology. The study of the human machine, including oxidizing nutrients and generating metabolic byproducts.

positive-negative-interesting. Using brainstorming techniques to develop the positive, negative, and interesting points to choose between several space plan options.

process charts. Show the type of activity that acts on a product.

process focus. A manufacturing focus that allows each department to specialize in their particular process or craft.

product focus. A manufacturing process where operations are grouped into departments that focus on products. Each department must have all operations equipment and skills, yet only process a single product.

product-volume analysis. Examines the current and future time frames for the products and their volumes so that the relationships between various products may be understood.

production flow analysis. Using spreadsheets, computer analysis, or manual methods to group parts that have similar routings.

project cell. A work cell that produces a wide variety of unrelated products using multiple processes.

proportioning. One of six methods used for calculating space requirements. The space for a given SPU comes from the calculation of another space.

ratio forecasting. A method of calculating space requirements, it uses historical trends to forecast space.

reorder point. A production maintain a significant stock of each item. As users withdraw items, the remaining stock declines. When the stock reaches a pre-determined level—the reorder point—a signal initiates a purchase order or shop order to replace the stock.

site saturation plan. A plan that places the maximum production space on the site. Also included are appropriate support spaces such as docks, parking, offices, and warehouses.

socio-technical system. The integration of social and technical systems into one unit. Social systems include people and their habitual attitudes, values, behavioral styles, and relationships. Technical systems include machinery, processes, procedures, and their physical arrangement (layout).

space plan primitive. An affinities diagram (q.v.) with space requirements added to it.

space plan. Determines and lays out the use of space in and around a facility, ranging from site planning to workstation design. Also known as a layout.

space planning units. The entities arranged by space plan designers. Also called cells and activity areas.

space standards. One of six methods used for calculating space requirements. Based on experience, space requirements that relate to some parameter of the business are established according to job classifications or places in the hierarchy.

static balance. When the equipment times for operations, averaged over the entire product mix, are similar or the same.

static work. Tasks that require an operator's body to remain in a stationary position for an extended period.

strategic layout planning. Top-down facilities planning that sets policy first and then arranges the technology, organization, and facility to support it.

stripe plan. A special version of a zone plan (q.v.) where each SPU expands horizontally.

sub-micro-space planning. The space plan level at which a workstation is designed; tasks, operators and tools are examined; tasks are allocated; and tools and fixtures are selected or designed.

supra-space planning. The space plan level at which site planning takes place. This includes number, size, and location of buildings, infrastructure, plant expansions, and eventual site saturation.

systematic layout planning. Using step-by-step procedures, conventions, and phases to plan a layout, adding system and structure to the plan.

terminal route structure. A route structure that works well with long distances and partial loads. Loads move first to several terminals, often on different equipment for each terminal, and then on to their destination.

transformation. One of six methods used for calculating space requirements. A designer takes a space requirement for an existing SPU and uses that as the basis for projecting future needs.

visual estimating. A method of calculating space needs. Templates representing equipment and furniture—are placed on a layout according to a designer's judgment and knowledge of similar installations.

weighted factor analysis. Decision-making based on a combination of the various factors, both qualitative and quantitative. Judges first identify the factors, decide a weight for each, and rate each option.

work cell operations plan. A form for recording infrastructure decisions.

work cells. Small, self-contained work units.

worker selection. The process of identifying characteristics for each job and the minimum criteria for those who might be selected for it.

working facilities. The land, buildings, and equipment that provide the physical capability to add value.

workstation. The space or area of a facility in which individuals or operators perform tasks. This ranges from an assembly station in a factory to a cube in an office.

zone plan. In site planning, a development plan that begins with a central core of space. Each SPU, function, or activity has a related zone. As space needs increase, the SPUs expand outward but within their designated zones.

Bibliography

Adair, John. 1987. *Effective teambuilding*. London: Pan Books.

Allegri, T. H., Sr. 1984. *Materials handling principles and practice*. New York: Van Nostrand Reinhold.

ANSI/HFES 100-1988. 1988. Human Factors and Ergonomics Society.

Apple, J. M. 1972. *Materials handling systems design*. New York: Ronald Press.

Baetz, Mary L. 1985: *The human imperative: Planning for people in the electronic office*. Toronto: Holt, Rinehart, and Winston.

Barnes, R. M. 1968. *Motion and time study*. New York: Wiley.

Bolz, Harold A. and George E. Hagemann. 1958. *Materials handling handbook*. New York: Ronald Press.

Boston technical furniture application handbook. 1994. 4-6.

Brandt, Peter B. 1992. *Office design*. New York: Watson-Guptil.

Buffa, Elwood S. 1984. *Meeting the competitive challenge*. Homewood, Ill.: Dow Jones-Irwin.

Bureau of labor statistics. 1993.

Doll, William J. and Mark A. Vonderembse. 1990. The evolution of manufacturing systems: Towards the post-industrial enterprise. *Proceedings of the 5th International Conference of the Operations Management Association*. Warwick, England: University of Warwick Press.

Duffy, Francis, Andrew Laing, and Vic Crisp. *The responsible work place: The redesign of work and offices*. Oxford, England: Butterworth-Heinemann.

Durand, Jacques. 1972. A new method for constructing scenarios. *Futures*. 325-330.

Eastman Kodak Co. 1986. *Ergonomic design for people at work*. New York: Van Nostrand Reinhold.

Eastman Kodak Co. 1983. *Ergonomic design for people at work*. 1: 15-17.

Ford, Henry. 1926. *Today and tomorrow*. Garden City, N.Y.: Doubleday.

Francis, Richard L. and John A. White. 1974. *Facility layout and location: An analytical approach*. Englewood Cliffs, N.Y.: Prentice-Hall.

Fraser, M. 1989. *The worker at work*. New York: Taylor and Francis.

Frederickson, Jack M. 1989. *Designing the cost effective office: A guide for facilities planners and managers*. New York: Quorum Books.

Friedman, Arnold, John F. Pile, and Forrest Wilson. 1982. *Interior design: An introduction to architectural interiors*. New York: Elsevier.

Haynes, D. O. 1958. *Materials handling applications*. Philadelphia: Chilton.

Hays, Robert H. and Steven C. Wheelwright. 1984. *Restoring our competitive edge*. New York: Wiley.

Hicks, Philip E. 1977. *Introduction to industrial engineering and management science*. New York: McGraw-Hill.

Hill, Terry. 1985. *Manufacturing strategy*. London; MacMillan Education.

Hodson, William K., ed. 1992. *Maynard's industrial engineering handbook*. 4th Edition. New York: McGraw-Hill.

Hudson, Kenneth. 1976. *The archaeology of industry*. New York: Charles Scribner's Sons.

Institute of Industrial Engineers. 1983. *Industrial engineering terminology*. Norcross, Ga.: Institute Of Industrial Engineers.

Knobec, Lance. 1987. *Office furniture*. New York: E. P. Dutton.

Konz, S. 1979. *Work design*. Columbus, Ohio: Grid Publishing.

Kroemer, H. K. E. and H. J. 1990. *Engineering physiology*. New York: Van Nostrand Reinhold.

Kulwiec, R. A. 1985. *Materials handling handbook*. 2nd edition. New York: Wiley-Interscience.

Lee, Quarterman. March 25, 1986. Layout for the just-in-time plant. *National Plant Engineering and Maintenance Conference Proceedings*. Chicago.

Lee, Quarterman. 1986. Quality and productivity trends. Latin American Quality Control Congress. Imecca, Mexico.

Lee, Quarterman. 1988. Computer-aided plant layout with real-time material flow evaluation. *Journal of Engineering Computing and Applications CAD/CAM Management Strategies*. 1:559.

Lee, Quarterman. June 1990. Manufacturing focus: A comprehensive view. *Manufacturing Strategy*. Technical and Medical Publishers. Christopher A. Voss, ed. London: Chapman and Hall Scientific.

Lee, Quarterman. October 1991. Integrated plan layout for just-in-time and world class manufacturing. *Engineering Management Conference Proceedings*. Orlando, FL.: JIE International.

Lee, Quarterman. November 19-20, 1991. Reconfiguring a warehouse for just-in-time operations. Presented at Society of Manufacturing Engineers. Schaumberg, IL.

Lee, Quarterman and M. Niedenthal. March 16, 1987. Layout for the just- in-time plant. *National Plant Engineering and Maintenance Conference Proceedings*. Chicago, IL.

Lee, Quarterman and W. Wrennall. October 28-31, 1990. Manufacturing focus for strategic advantage. *JIE Intergrated Systems Conference and Society for Integrated Manufacturing and Conference Proceedings*. San Antonio, TX.

Liberman, Marvin B. Summer 1987. Strategies for capacity expansion. *Sloan Management Review*. 28:4.

Lueder, Kageyu. 1994. *Hard facts about soft machines*. New York: Taylor and Francis.

Lynch, K. 1992. Site Planning. Cambridge, Mass.: MIT Press.

Material handling engineering 1990–1991 handbook and directory. 1990. Cleveland, Ohio: Penton Publishing. 44:13.

McCormick, Ernest J. and Mark S. Sanders. 1987. *Human factors in engineering and Design*. New York: McGraw-Hill.

Miller, Stanley S. 1988. *Competitive manufacturing*. New York: Van Reinhold.

Mital, Nicholson Ayoub. 1993. *A guide to manual material handling*. New York: Taylor and Francis.

Morris, W. T. 1962. *Analysis for materials handling management*. Homewood, Ill: Richard D. Irwin.

Muller, Thomas. 1983. *Automated guided vehicles*. New York: IFS Publications.

Muther, Richard. 1973. *Systematic layout planning*. 2nd edition. Boston: CBI Publishing.

Muther, Richard and Knut Haganas. 1969. *Systematic handling analysis*. Kansas City: Management and Industrial Research Publications.

Muther, Richard and Lee Hales. 1980. *Systematic planning of industrial facilities*. Vol. 1 and 2. Kansas City: Management and Industrial Research Publications.

Nance, Harold W. 1983. *Office work measurement*. Malabar, Fla.: Robert F Kreiger.

Neibel, B. W. 1972. *Motion and time study*. Homewood, Ill: Irwin.

Pelmear, Taylor, Wasserman. 1992. *Hand-arm vibration*. New York: Van Nostrand Reinhold.

Pheasant, S. 1988. *Bodyspace*. New York: Taylor and Francis.

Porter, Michael E. 1980. *The competitive strategy*. New York: Free Press.

Pressman, Andy and Peter. 1980. *Integrated space systems vocabulary for room language*. New York: Free Press.

Priest, Robert. 1968. *The office: A facility based on change*. Elmhurst, Ill.: Business Press.

Pulat, B. Mustafa and David Alexander. 1991. *Industrial ergonomics: Case studies*. Norcross, Ga.: Institute of Industrial Engineers.

Putz-Anderson, V. 1988. *Cumulative trauma disorders*. New York: Taylor and Francis.

Quin, Karen Takle. 1985. *Advances in office automation*. Chicester: John Wiley and Sons.

Ray, Darrel and Howard Bronstein. 1995. *Teaming up*. New York: McGraw-Hill.

Salvendy, Gavriel. 1987. *Handbook of human factors*. New York: Wiley.

Salvendy, Gavriel, ed. 1992. *Handbook of industrial engineering*. 2nd edition. New York: Wiley-Interscience.

Saphier, Michael. 1978. *Planning the new office*. New York: McGraw-Hill.

Schaefer, G. 1988. *Functional analysis of office requirements: A multiperspective approach*. Chicester: John Wiley and Sons.

Scholtes, Peter. 1988. *The team handbook*. Madison, Wis.: Joiner Associates.

Senge, Peter M. 1990. *The fifth discipline*. New York: Doubleday.

Shinohara, Isao. 1988. *New production system*. Cambridge, Mass: Productivity Press.

Skinner, Wickham. 1985. *Manufacturing: The formidable competitive weapon*. New York: Wiley.

Steele, Fritz. 1986. *Making and managing high quality workplaces: An organizational ecology*. New York: Teachers' College Press.

Stibic, Vladimir. 1982. *Techniques and methods for intellectual work*. Amsterdam: North-Holland Publishing.

Szenasy, Susan S. 1984. *The office book design series: Private and executive offices*. New York: Facts on File.

Tomkins, James A. and J. A. White. 1984. *Facilities planning*. New York: Wiley.

Tomkins, James A. and J.D. Smith. 1988. *The warehouse management handbook*. New York: McGraw-Hill.

Tweedy, Donald B. 1985. *Office space planning and management*. New York: Quorum Books.

Voss, Christopher A., ed. 1992. *Manufacturing strategy: process and content*. London: Chapman and Hall.

Wack, P. 1985. Scenarios: Uncharted waters ahead. *Harvard Business Review*. 5: 73-89.

Wack, P. 1985. Scenarios: Shooting the rapids. *Harvard Business Review*. 6: 73-89.

White, Anthony. 1971. *Management objectives in office planning and implementation*. London: Industrial and Commercial Techniques.

White, J. A. 1987. *Production handbook*. 4th edition. New York: Wiley.

Wineman, Jean D. 1986. *Behavioral issues in office design*. New York: Van Nostrand Reinhold.

Wrennall, W. and Quarterman Lee. 1989. Achieving requisite manufacturing simplicity. *Manufacturing Technology International*. London: Sterling Publications.

Wrennall, W. and Quarterman Lee. 1994. *Handbook of commercial and industrial facilities management*. New York: McGraw-Hill.

Woodson, Wesley E. and Donald W. Conover. 1966. *Human engineering guide for equipment designers*. 2nd rev. edition. Berkeley: University of California Press.

Index

V

W

Z

About the Authors

Quarterman Lee is the president of Strategos, Inc., a consulting firm. He was an editor of the *Handbook of Commercial and Industrial Facilities Management* and is the author of numerous articles in the fields of industrial management and engineering. He has created training courses for the Institute of Industrial Engineers, the Society of Manufacturing Engineers and the U.S. Postal Service and assisted organizations worldwide with the application of innovative management methods.

Arild Amundsen is president of Plancraft, AS, and has an extensive background in mechanical engineering, consulting, and management. He resides in Oslo. **William Nelson** is president of Humatech, Inc., and specializes in bio-socio-technical systems. He also designs products, industrial equipment and workstations. **Herbert Tuttle** is an assistant professor in the Graduate Engineering Management Program at the University of Kansas. He has over 15 years of experience in industrial engineering and plant management.

About EMP

ENGINEERING & MANAGEMENT PRESS (EMP) is the book publishing division of the Institute of industrial Engineers. EMP was founded in 1981 as Industrial Engineering and Management Press (IE&MP). In 1995, IE&MP was "reengineered" as ENGINEERING & MANAGEMENT PRESS by an entirely new staff.

As both IE&MP and EMP, the press has a history of publishing successful titles. Recent successes include: *Toyota Production System, 2nd Edition; Beyond the Basics of Reengineering; Business Process Reengineering: Current Issues and Applications;* and *Managing Qulaity in Americas Most Admired Companies.*

EMP's newest titles are *By What Method?; Simulation Made Easy;* and *Essential Career Skills for Engineers,* the first in the **Engineers in Business Series,** of which this book is a part.

For more information about EMP or to request a free catalog of EMP's current titles, please call IIE Member & Customer Service at 800-494-0460 or 770-449-0460.